Building a Better Race

Building
a Better Race

Gender, Sexuality, and Eugenics
from the Turn of the Century
to the Baby Boom

Wendy Kline

UNIVERSITY OF CALIFORNIA PRESS

Berkeley / Los Angeles / London

University of California Press
Berkeley and Los Angeles, California

University of California Press, Ltd.
London, England

First paperback printing 2005
© 2001 by the Regents of the University of California

Library of Congress Cataloging-in-Publication Data

Kline, Wendy.
    Building a better race : gender, sexuality, and eugenics from the turn
of the century to the baby boom / Wendy Kline.
        p.   cm.
    Includes bibliographical references and index.
    ISBN 0-520-24674-8 (pbk: alk. paper)
    1. Eugenics—United States—History.   2. United States—Moral
conditions.   I. Title.
HQ755.5U5 K55   2001
363.9'2—dc21   2001027246

Manufactured in the United States of America

14   13   12   11   10   09   08   07   06   05
10   9   8   7   6   5   4   3   2   1

To my teachers
Ellis Turner,
Dan and Helen Horowitz,
and Karen Halttunen,
who connected me to the past,

and

to my son,
Max William Paula,
born December 14, 1999,
who connects me to the future

Some people find in the word "eugenics" only a subject for humor. But if you are yourself a eugenically superior person, you realize that the ideal of sound minds and sound bodies, which is the core of the eugenics movement, is not only admirable but necessary. If the human race, beset by the growing complexities of civilization, is to endure and be able to enjoy life in the future, it must free itself from the physical and mental handicaps which now beset it.

Paul Popenoe and Roswell Johnson,
flyer for *Applied Eugenics,* 1933 edition

Contents

Illustrations

Acknowledgments

Without the encouragement, support, and assistance of several people, this project never would have made it into print. I have been blessed with excellent teachers since high school who instilled in me a love for learning and specifically for history—this book is dedicated to them. I would like to thank in particular Gail Jackson and Ellis Turner at Sidwell Friends. At Brown University, Mari Jo Buhle's exceptional teaching drew me to the field of women's history. Teaching assistant Gail Bederman had a passion for the subject that was infectious, and her work, *Manliness and Civilization*, influenced the direction of my own dissertation. At Smith College, Dan and Helen Horowitz taught me that becoming a historian is not only a worthy pursuit but one that is incredibly fulfilling. They continued to encourage and advise me long after my days at Smith were over. I initially discovered the delight of sorting through archival materials while working for Helen on her own research project, *The Power and Passion of M. Carey Thomas*. She also read my dissertation closely and offered helpful advice on how to turn it into a book. At the University of California, Davis, several teachers and graduate students made my experience extremely rewarding, including Jill Hough, Rosamaria Tanghetti, Alan Taylor, Mike Dietrich, and Cathy Kudlick. But I owe the greatest debt to Karen Halttunen, my dissertation adviser. Her high standards, incisive questions, and unflagging interest improved my project at every stage of its development. Her scholarly expertise and her commitment to excellence are a source of inspiration to me.

My research and writing would not have been possible without financial assistance. I am grateful to the Woodrow Wilson and Johnson & Johnson Foundations for a dissertation grant in women's health, as well as to the California Institute of Technology Archives and the Social Welfare History Archives at the University of Minnesota for research grants. Various sources at the University of California, Davis, provided me with financial assistance, including the history department, the Pro Femina Research Consortium, and the Humanities Institute. I would also like to thank the University Research Council at the University of Cincinnati for financial support.

Friends and family also helped me throughout my research and writing, providing me with a bed or a warm meal while on research trips or with words of encouragement: Susan Bell, Bruce Silva and Tim Barkalow, Paul and Michelle Gardner, Andrea Friedman, Ajoy Chakrabarti, Nancy Kline and Christopher Spence, Jennifer Lloyd, Maureen Kline, and Barbara and Dick Gardner. I am also grateful to archivists Shelley Erwin at the California Institute of Technology Archives and David Klaassen at the Social Welfare History Archives for their tireless assistance both during my research visits and through e-mail.

Living in Munich, Germany, made revising the dissertation into a book even more of a challenge than it otherwise would have been. A number of people helped make this possible, particularly Monica McCormick at the University of California Press, whose enthusiasm and support for the project brought it to completion. I would also like to thank the two scholars who read my manuscript for the Press, particularly Regina Morantz-Sanchez. Her insightful comments and suggestions enabled me to approach my work from a fresh perspective and to rethink the more problematic parts of the manuscript. I am also indebted to Pamela Fischer for her thoughtful and thorough copyediting of the manuscript. Jennifer Terry became my "Internet angel." After I contacted her via e-mail with a research question, she generously shared material and insights from her book (then forthcoming), *An American Obsession: Science, Medicine, and the Place of Homosexuality in Modern Society.* Her frequent friendly and informative messages kept me going when the isolation of working alone and abroad threatened to slow me down. Tiffany Wayne tracked down sources and mailed me material that I could not get in Germany, never complaining as I bombarded her with questions from afar.

My time in Germany proved fruitful in more than one way: On December 14, 1999, I gave birth to my first child, Max William Paula. I watched the manuscript grow along with my belly, completing my revisions just weeks before his arrival. Having my own child reminded me that the drama of eugenics and its quest for reproductive morality affected Americans in a personal, as well as political, manner.

Though I disagree with eugenicists' reasons for stressing the importance of the family (as well as how they defined it), I too have found it a source that I could not do without. Thank you to my own little family: to Max and to my husband, Stefan, who bore the brunt of the stress of my simultaneously producing book and child and who rarely complained about it, instead offering me love and encouragement. Though most of the time it was cloudy outside during our stay in Munich, he made sure that there was plenty of sunshine inside.

Introduction

"The American woman is the leader of the awakened social conscience in a country-wide crusade that is cooperating to build a better race," declared Progressive reformer Mabel Potter Daggett in 1912.[1] Between 1900 and 1960, this "country-wide crusade" to strengthen family and civilization by regulating fertility—more commonly known as eugenics—developed into a powerful and popular ideal. The extraordinary story of how this crusade came about and why Americans found it appealing is noticeably absent from most studies of American culture. Yet if we continue to approach eugenics as merely an embarrassing mistake with little historical significance, we will never understand the movement's powerful appeal to generations of Americans concerned about the future of morality and civilization.

In the early twentieth century, what Sigmund Freud called the "civilized morality" of the white middle class began to lose ground. Social observers noted the decaying morality of American youth, particularly working-class girls who refused to abide by genteel standards of Victorian femininity. By the 1920s, "civilized morality" had been replaced by what enthusiasts termed the "new morality." One female supporter announced triumphantly in 1924 the "first result of the new morality": "women are demanding a reality in their relations with men that heretofore has been lacking, and they refuse longer to cater to the traditional notions of them created by men." Whereas "civilized morality" promoted female passionlessness and public reticence about sex, the "new morality" celebrated an equality of desire between the sexes. "The myth of the pure woman is at an end," declared an enthusiast. As a popular magazine announced on the eve of World War I, it was "sex o'clock in America."[2]

This revolution in manners generated profound concerns about the future of morals in America. Traditional morality was a "dreary" concept, one proponent of the "new morality" observed.[3] Associated with female purity and passionlessness, it did not make sense in a modern world. Yet many found that the "new morality" of the 1920s was not an adequate replacement for traditional morality. They feared the consequences of modern-day moral disorder (exemplified in the sexual freedom and expressiveness of the "new morality") and sought a successor that would stabilize rather than threaten marriage and family. The "true woman" had been the moral guardian of civilization in the nineteenth century; who would be her successor in the twentieth?

In this book, I seek to explain how and why eugenics became an appealing solution to the problem of moral disorder. Eugenics elicited tremendous popular and professional support, I argue, because it linked two issues of great concern to the white middle class in early-twentieth-century America: race and gender. Social and economic changes threatened to undermine established race and gender hierarchies. The fear that "race suicide" would result from the fact that the birthrate of the American-born white middle class was dropping well below that of immigrants and the working class heightened public alarm over this loss of authority. Unless fertility was regulated to compensate for the differential birthrate ("more children from the fit, less from the unfit," in Margaret Sanger's words), moral and racial decay would bring the downfall of civilization.[4]

Applying eugenic principles to modern civilization, many came to believe, would regulate fertility and thus counter this decline. Specifically, eugenicists wanted to make motherhood an exclusive privilege rather than an inherent right by encouraging the "fit" (primarily the white middle class) to have more children while restricting the "unfit" from doing so. They sought to modernize morality by casting it as a racial and reproductive concept. Americans could build a better race, they believed, by instilling a sense of "reproductive morality" into the public consciousness. Reproductive decisions would then be based not on individual desire but on racial duty. Tomorrow's children needed sound minds and healthy bodies in order to strengthen civilization. Therefore, reproduction should be limited to adults exhibiting these traits. Everyone, even those sterilized, eugenicists argued, would benefit from this moral system, which guaranteed a secure future for the race. Eugenics thereby offered an efficient moral system to replace nineteenth-century "civilized morality."

This book traces the emergence and evolution of eugenics from a movement focused primarily on preventing procreation of the "unfit" ("negative eugenics") to one centered on promoting procreation of the "fit" ("positive eugenics"). Chapter 1 analyzes social concerns about race, gender, and middle-class authority between 1890 and 1915 in order to explain why and how eugenics became a central ideology and movement of the twentieth century. As the white middle-class birthrate began a precipitous decline, eugenicists commented crassly that the "morons" were "multiplying like rabbits."[5] Eugenicists promoted two opposing models of womanhood that suggested the importance of gender to eugenic ideology: the "mother of tomorrow" and the "moron." The mother of tomorrow represented the procreative potential of white middle-class womanhood, while the moron symbolized the danger of female sexuality unleashed. Together, these models, which carried great symbolic weight in the eugenics movement, demonstrated that the eugenic definition of womanhood was double-edged: it portrayed women as responsible not only for racial progress but also for racial destruction.

Chapter 2 addresses the negative-eugenics campaign by analyzing its impact on the institutional level. The chapter focuses on the history of the Sonoma State Home for the Feebleminded from its founding in 1884 to the 1930s, by which time it had become the fastest-growing public institution in California and a nationwide leader in the number of eugenic sterilizations performed. Initially, eugenicists believed that quarantining the female "high-grade moron" would prevent sexually promiscuous women from infecting the race. But by the 1910s, promiscuous sexual behavior had spread into the middle classes. Segregating the promiscuous was no longer an adequate or cost-effective strategy. Instead, sterilization gained popularity as an efficient way to prevent the spread of mental and moral deficiency to future generations. Segregation and sterilization at Sonoma set the stage for a new popular interest in eugenics. It provided proof that motherhood could—and should, in the minds of many—be restricted.

Chapter 3 addresses the strategies implemented to gain the support of organized medicine and the masses at a time when the act of preventing conception smacked of obscenity. Many professionals, particularly doctors, were opposed to regulating reproduction because they believed it might undermine their credibility. Eugenicists had to convince organized medicine that sterilization was a legitimate, effective, and ethical procedure for preventing procreation in order to improve the race. Portraying the procedure as a harmless and effective remedy, sterilization advocates enabled the eugenic-sterilization campaign to take center stage in medical and popular debates about female sexuality and reproduction. By framing female reproductive behavior in the scientific language of eugenics, they helped to shape the development of modern sexual ideology and reproduction on their own terms. Their campaign also contributed to a dramatic increase in the number of sterilizations performed in the United States. By the 1930s, the annual average number was 2,273 per year—almost ten times the average between 1907 and 1920.[6]

Chapter 4 explores a significant strategic shift in the eugenics movement that enabled it to prosper long after many historians claim it had disappeared. Because eugenicists were being attacked for their emphasis on heredity as the primary justification for sterilization, they incorporated an environmental explanation as well. Specifically, they focused on the importance of motherhood and family to the future of the race. If, as social scientists now argued, a mother shaped her child's development, then it was all the more important to ensure that only "fit" women became mothers. Eugenicists thereby used the new emphasis on environment to their advantage. A sensational court case involving the coerced sterilization of twenty-year-old Ann Cooper Hewitt, heiress to her late father's millions, signaled this new phase in eugenic strategies. The central issue was not whether she suffered a hereditary deficiency but whether she would make a desirable mother. The case, debated by legal experts, sensationalized by journalists, and followed by millions, introduced sterilization as a family-centered solution to the problem of female sexuality.

With motherhood and family preservation at the center of their campaign beginning in the 1930s, many eugenicists turned to positive eugenics, or promoting the procreation of the eugenically fit, as an additional strategy. Chapter 5 analyzes the influential role of eugenicists in marriage and family counseling between 1930 and 1960. By educating young adults about the principles of selecting a eugenically fit mate and by counseling those already married on how to preserve and protect their marriage and family, Robert Dickinson, Lewis Terman, and Paul Popenoe hoped to alter the way ordinary Americans thought about their own decisions regarding marriage and procreation.

This campaign yielded the greatest success for the eugenics movement. The shift in the 1930s to a more environmental approach also permitted positive eugenics to emerge as a popular movement in the 1940s and 1950s. Eugenicists' plea for white middle-class women to become more "family-minded" was answered in the pronatalism of the 1950s. Between 1900 and 1960, eugenicists appealed to a wide audience of Americans concerned about moral disorder and interested in building a better race. Initially focused on preventing the procreation of the unfit, they also succeeded in persuading middle-class Americans to reconsider their procreative potential.

In the 1990s a number of significant works were published that demanded a reevaluation of the eugenics movement. Studies of eugenics in Denmark, Sweden, Norway, Brazil, Latin America, Russia, and the American South brought a variety of voices and new meanings to the history of eugenics in the twentieth century and challenged the notion that eugenics was a unilateral conservative social movement. Other studies teased multiple meanings out of the movement by looking at mass culture and the diversity of support eugenics received. Ian Dowbiggin, for example, characterizes the movement as inclusive and pluralist, concluding that "it meant different things to different people in different settings."[7]

These new studies helped to generate increased interest in the history of eugenics. Yet there is more work to be done. One major problem is that most histories of eugenics pay little attention to gender.[8] Though several historians recognize the importance of sexuality and gender to the eugenics movement, they limit their analysis to the role of women in the movement. Indeed, too often, *gender* has merely become a synonym for *women*.

The role of gender in the eugenics movement requires complex analysis. Gender was important because women played "a significant part in the politics of eugenics," which gave them "a new space for social action," as scholars now argue.[9] But gender was also central to eugenics because the movement called for a new approach to understanding sexuality, reproduction, and the role of men and women in society. Some women actively supported eugenics; some, as physicians, even sterilized other women; still others lobbied against eugenics. They did not form a unitary coalition, but this lack of unity does not make gender any less relevant to our understanding of eugenics, as I will demonstrate in the pages that follow.

Race also needs clarification here. American eugenicists often made references to "improving the race" without specifying whether they meant the human race, the Anglo-Saxon race, or some other type of racial differentiation. Though the meaning varied from person to person, the common eugenic vision of "building a better race" was implicitly racist. While most of those targeted as in need of sterilization in California's movement were white, race was still at the center of the eugenic campaign. As Hazel Carby writes in *Reconstructing Womanhood,* "Work that uses race as a central category does not necessarily need to be about black [people]." Eugenicists were agents of a racial discourse, and though it was sometimes unmentioned or merely hinted at, race was a salient category in the eugenics movement.[10]

This book traces anxieties about race and gender as expressed in eugenics from the turn of the century to 1960. My decision to bring eugenics into the 1950s is significant, for it challenges the historiography and the assumption that "mainline" eugenics died out well before World War II only to reemerge in a reformed state in the most recent past. Looking closely at the meaning of eugenics in the 1930s and 1940s, we can see clearly that continuity, rather than discontinuity, characterizes the ideas and actions of eugenicists at mid-century. Given the atrocities of Nazi eugenics (and the links between the German and American programs), it becomes difficult to fathom why many eugenic promoters continued their campaign long after the Nazi campaign was revealed. But American eugenicists, like many German eugenicists prior to National Socialism, did not identify with the Nazi version of eugenics and therefore saw no reason to alter their own eugenic goals even after 1945 (though they were careful to monitor their language, avoiding explicit racial comments that could associate them with the Nazis).[11] To fully understand the impact of eugenics on American culture, it is imperative that we incorporate the eugenic ideals of postwar America, which had a significant impact on gender and the pronatalism of the 1950s. One of my major goals in this book, then, is to place eugenics at the center of modern reevaluations of female sexuality and morality. I hope in this way to contribute to a new understanding of the influence of eugenics on twentieth-century American culture.

CHAPTER ONE

Motherhood, Morality, and the "Moron"

The Emergence of Eugenics in America

In 1915, despite the onset of World War I and the lingering effects of the 1906 earthquake, San Francisco hosted the Panama Pacific International Exposition. Recovering from destruction at home and fearing greater destruction abroad from war, Americans flocked to the Exposition to witness the celebration of progress, science, faith, and enlightenment as embodied in the architecture, exhibits, and dramas of the fair. "An international exposition is a symbol of world progress," one guidebook to the Exposition explained. "This one is so complete in its significance, so inclusive of all the best that man has done, that it is something more than a memorial of another event. It is itself epochal."[1]

This theme of progress, which drew on the recent completion of the Panama Canal, resounded in the sculptures depicting a new link between world cultures. At the entrance to the Exposition, close to the "column of progress," a giant group of sculptures mounted on a triumphal arch depicted the "Nations of the West," and it faced an equally dramatic "Nations of the East." Rising 160 feet in the air, the sculptures could "be seen at all points in silhouette against the sky." At the center of the cluster of Westerners, which included a French-Canadian trapper, a Latin American, a German, and an Italian, stood a pioneer woman called "the Mother of Tomorrow." Rising above her, a female spirit of "Enterprise" served "as a challenge to future generations to take up the pioneer spirit."[2]

Though this focus on progress at the 1915 exposition was not new, its gendered representation was. America's fairs had previously celebrated the advancement of civilization, most notably in the construction of the

7

White City at the 1893 Columbian Exposition in Chicago, which Congress authorized as "an exhibition of the progress of civilization in the New World." As historians have shown, the White City "depicted the millennial advancement of white civilization" as exclusively male.[3] Much to the dismay of the Exposition's Board of Lady Managers, women's contributions to progress received little acknowledgment at the 1893 Exposition. In late-nineteenth-century America, the discourse of civilization underscored the superiority of white manhood and suggested that white men, not women, would lead the way toward human perfection.

But by 1915 white womanhood was considered no longer a marginal player in the evolution of the race but a central figure. As the "mother of tomorrow," she controlled the racial makeup of future generations. Her larger-than-life presence symbolized the widespread cultural debate on the meaning of womanhood in the twentieth century. Specifically, by the early twentieth century, white female fertility had become a potential panacea for the problems of racial degeneracy. As feminist socialist Charlotte Perkins Gilman explained in *Women and Economics* (1899), women of the new century needed to recognize not only "their social responsibility as individuals, but their measureless racial importance as makers of men."[4]

The Rise of Eugenics as a Central Ideology

What accounts for this new emphasis on white motherhood as the key to racial progress in the early twentieth century? In order to understand the significance of this shift, we need to investigate the meaning of "racial progress" at the turn of the century in the United States and, more generally, the increasing popularity of eugenics as an ideal Progressive reform. How and why did eugenic ideology gain authority as a scientific solution to the perceived problem of racial degeneracy—the primary threat to racial progress? Looking at social concerns about race, gender, and middle-class authority between 1890 and 1915 will help answer this question.

During the 1890s, a combination of factors led to an "unprecedented outpouring of racism" in the middle class. Economic changes accompanying the growth of corporate capitalism threatened to annihilate the nineteenth-century "cult of the self-made man"; between 1870 and 1910, the percentage of self-employed workers dropped from 67 to 37 percent. By 1900, the middle-class economy had become corporatized and bureaucratized. The future "lay not in fields and shops, but inside the walls of enormous corporations." As a result, men lost touch with the product of their work.[5] Severe economic depression and labor unrest—thirty-seven thousand strikes in two decades—increased anxieties that the middle class was losing social authority. As Alan Trachtenberg suggests, the growth of corporate power "wrenched American society from the moorings of familiar values," and the result was a widespread sense of lost autonomy.[6]

Echoing these external, economic factors that threatened the power of the white middle class and, in particular, white manhood, internal symptoms of increasing fragility and weakness manifested themselves in the male body. Late-nineteenth-century doctors discovered a new disease of the middle class: neurasthenia. George Beard, who discovered neurasthenia, defined it as "a lack of nerve force."[7] Wrestling with the challenges of corporate bureaucracy, middle-class men had to exercise "excessive brain work" in order to succeed. As popular novelists such as Henry James and William Dean Howells noted in the late nineteenth century, the ideal of economic and social advancement was a "major cause of social dislocation and individual unhappiness."[8] It often resulted in nervous strain, which, combined with Victorian social pressures for sexual self-containment, drove more and more men to mental and physical illness. To many commentators, the outbreak of neurasthenia was a sign that white middle-class manhood had lost its virility.[9]

From the perspective of the white middle class, African American men, in contrast, appeared to possess powerful masculinity. The majority of whites in both the North and the South, for example, supported the growing practice of lynching in the South as a justification for the actions of the "negro rapist." The virility of the African American male was in fact so powerful, argued supporters of lynching, that it needed to be restrained. It threatened the sanctity of southern white womanhood and its emphasis on chastity.[10]

While the white middle class appeared to be fading away, the strength and numbers not only of African Americans but also of the working class and immigrants seemed to increase. The popularity of pugilism and other combative male sports in working-class culture suggested to middle-class whites that working-class manhood had not lost its virility.[11] In addition, immigration from southeastern Europe soared during what John Higham calls the "nationalist nineties."[12] As Frederick Jackson Turner was announcing the "closing of the American frontier," the influx of immigrants meant that nearly 15 percent of America's population in 1890 was foreign-born.[13] Working-class and immigrant men seemed to possess a potency that white middle-class men, weakened by corporate jobs and mental strain, had lost.

Also threatening the authority of white middle-class manhood in the late nineteenth century was the "woman question." Turn-of-the-century social observers frequently commented on two new class-based figures in American culture: the "woman adrift" and the "new woman." Both figures called into question the sanctity of nineteenth-century gender roles and signaled the end of the Victorian era.

The working-class "woman adrift" appeared on the urban landscape at the end of the nineteenth century. New economic opportunities in retail and industry drew young single women into cities, where they lived apart from their families. In 1900, one in five urban working women lived on her own.[14] Living and working independently outside the domestic sphere, these women challenged nineteenth-century middle-class conceptions of womanhood. Not surprisingly, their presence generated a great deal of concern, and, in particular, their sexuality became the focus of public anxiety.[15] Frequenting dance halls and other new forms of commercialized leisure, these "charity girls" exhibited sexual behavior that was more relaxed than that of both earlier generations of women and modern middle-class women. They participated in a sexual culture in which sexual favors were exchanged for "treating," or gifts from men.[16]

A new generation of middle-class women also provoked consternation. The "new woman" challenged the existing social order by demanding "rights and privileges customarily accorded only to white middle-class men." Between 1890 and 1910, the number of women attending college tripled, and by 1920 women accounted for nearly 50 percent of the university population. Choosing college and careers over marriage and motherhood, these women violated the nineteenth-century "cult of true womanhood," which placed middle-class women in the home. As a result, they raised questions about gender and morality.[17]

Specifically, they generated concerns about the meaning of morality in the modern world. The nineteenth-century ideology of "separate spheres" ensured a balance between the corrupt and competitive male-oriented public sphere and the unadulterated, nurturing, female-oriented private sphere. As mothers and wives untainted by politics or public life, nineteenth-century middle-class women were believed to be the moral guardians of virtue. Thus when the new women of the late nineteenth and early twentieth centuries challenged the sanctity of "separate spheres" by attending college and demanding a public voice, they encountered a great deal of hostility. Anna Rogers, writing for the *Atlantic Monthly,* declared that "the devouring ego in the 'new woman'" had created "the latter-day cult of individualism; the worship of the brazen calf of Self."[18]

Because they challenged the conventional standards of womanhood, these women were attacked as "unsexed" or "mannish." Their demand for equal rights called into question the sanctity of gender roles. Many men believed that economic equality threatened to merge the two sexes into "dangerous confusion." Women were becoming masculine just as men were becoming increasingly weak and effeminate. Home and family were the cornerstones of society, and if women abdicated their domestic duties, what was to become of moral order? The new woman, one historian points out, was "the enemy of marriage, the home, and therefore civilization."[19]

Thus, as the growing presence and perceived virility of African Americans, immigrants, and the working class—as well as the increasing visibility of working-class "women adrift"—threatened white middle-class male authority in both power and numbers, proponents of racial progress targeted another factor in middle-class decline: the limited fecundity of this new woman. As Teddy Roosevelt proclaimed in the 1900s, white middle-class womanhood had willfully abandoned its fertility. The white birthrate was rapidly declining: whereas the average American family of 1840 had produced six children, that of 1900 generated only three.[20] Roosevelt propelled sociologist Edward Ross's term *race suicide* into the public arena. In a 1901 address, "The Causes of Race Superiority," Ross warned that the advancement and progress of the "superior race" could lead to its demise; manhood had become overcivilized, decadent, and impotent. But Roosevelt, significantly, placed the blame on white womanhood. Women of "good stock" who chose not to have children, he declared, were "race criminals."[21] Juxtaposing his imperialist foreign policy with an attack on race suicide at home, he argued that "no race has any chance to win a great place [in the world] unless it consists of good breeders as well as of good fighters."[22]

Linking fertility to war, Roosevelt raised the stakes of white middle-class procreation. The white race not only risked losing power but faced eventual extinction if women did not heed Gilman's call to "social responsibility . . . as makers of men." Roosevelt's use of "race suicide" generated a great deal of attention in the media; for example, between 1903 and 1905, *Popular Science Monthly* published sixteen articles and letters on the issue. "It is surely a serious problem," the publication declared, "when the more civilized races tend not to reproduce themselves."[23] In addition, more than thirty-five articles appeared in popular magazines between 1905 and 1909 discussing the infertility of white Americans.[24]

Though Congress had outlawed the dissemination of birth-control information and the American Medical Association had succeeded in criminalizing abortion by the late nineteenth century, middle-class white women still managed to curtail conception. In addition, they were marrying later (while the median age at marriage in 1860 was twenty-one, it was twenty-four by 1890), and many were choosing not to marry at all.[25] As educational and professional opportunities expanded in the late nineteenth century, these women found more to life than marriage and motherhood, and they restricted procreation accordingly. As one female journalist remarked at a roundtable discussion of the birthrate decline, "I think it is ridiculous for you men to sit here and say the things you have said tonight about a woman's duty to have children. Many of us have found other interests and compensations very absorbing."[26]

Those who shared Roosevelt's concern about race suicide blamed the new woman for neglecting her responsibility to be (in Gilman's words) a "maker of men." As one concerned scientist declared during a speech at the University of California, Berkeley, women were "destroying civilization" by turning away "from their natural modes of expression in the home and family." In their selfish attempts to "fill men's places," he continued, "they are sapping the strength of the race."[27]

By presenting motherhood in terms of "race progress," concerned commentators hoped that new women would see the error of their ways and return to home and hearth. Concerned about the future of the white race, as well as the advancement of civilization, Progressive reformers became interested in remaking womanhood in racial and millennial terms. As Gail Bederman explains, the obsession with the "advancement of civilization" beginning in the late nineteenth century stemmed from a millennial vision of "perfected racial evolution and gender specialization." Through intelligent procreation would come a better, more prosperous race of human beings.[28]

This millennial vision of progress, coupled with the desire to regulate fertility and stem the tide of "race suicide," exemplifies the new social concerns of the Progressive era. As Gilman declared, "The duty of human life is progress, development." Emphasizing the importance of social responsibility, she continued, "We are here, not merely to live, but to grow,—not to be content with lean savagery or fat barbarism or sordid civilization, but to toil on through the centuries, and build up the ever-nobler forms of life toward which social evolution tends."[29] In order to stem the tide of race suicide and to foster social evolution, greater regulation of fertility was needed.

These issues—social responsibility and state regulation—resonated in Progressive ideology. Progressivism represented a widespread and varied response to the multitude of changes brought by industrial capitalism and urban growth in the late nineteenth century. What drew these reformers together—from labor activists to clubwomen—was the desire for state intervention and regulation of social problems. Since the publication of Daniel Rogers's 1982 article "In Search of Progressivism," scholars have struggled to define and clarify the varied categories that constitute "Progressivism." Though it remains an "embattled concept," Progressivism is still an essential, albeit challenging, ideology that helps to explain a complex era in United States history.[30]

Making use of a new language of social efficiency and technical expertise, Progressive-era reformers approached social problems differently than their nineteenth-century forebears. In search of order in an increasingly complex world, they called for a new "social consciousness" to strengthen American civilization.[31] Within this context, the American eugenics movement emerged as a significant force that not only would shape immigration policy and sexual conduct during the Progressive era but would outlive its Progressive roots to gain increasing authority in the 1930s, 1940s, and 1950s.[32]

Like Progressivism, eugenic ideology comprised a complex combination of popular and scientific beliefs and interests that has confounded historians. As Diane Paul notes, "There is [currently] no consensus on what eugenics is." When the term was coined in 1883 by Francis Galton, a British statistician, he took the word from a Greek root meaning "good in birth." He defined eugenics as the science of improving human stock by giving "the more suitable races or strains of blood a better chance of prevailing speedily over the less suitable."[33] After studying prominent British families, he reasoned that most moral and mental traits, such as courage, intellect, and vigor, were passed on to offspring. Yet he also observed that the worthiest families produced the fewest children and believed that disaster would result if the situation were not reversed. He proposed that "those highest in civic worth" should be encouraged to have more children (defined as "positive eugenics") and those unworthy should be encouraged to have fewer or none ("negative eugenics").[34]

But in Progressive-era America, eugenic ideology appealed to reformers representing a wide range of interests and politics, who applied their own varied definitions of eugenics. Social radicals such as Gilman and Margaret Sanger embraced it as a civilizing force that would further the rights of women, as well as improve the race. Conservatives such as

Madison Grant (author of *The Passing of the Great Race*) viewed it as a justification for restraining the liberties of immigrants and the procreative powers of sexually promiscuous women. What they had in common was a vision of the future in which reproductive decisions were made in the name of building a better race, though they may have disagreed on how to go about achieving this goal. Indeed, one of the strengths of the eugenics movement was its widespread popular appeal to a diverse audience, which was due in large part to its decidedly vague definition. It is important that the words and actions of a few eugenic leaders not be allowed to define the parameters of eugenic meaning in America. Instead, as Nancy Stepan suggests, "we need to recapture 'ordinary' eugenics and its social meanings."[35]

Eugenics elicited tremendous popular and professional support because it linked two issues of great concern to the white middle class in early-twentieth-century America: race and gender. White middle-class authority and middle-class manhood both were in jeopardy because of social and economic changes that undermined established race and gender hierarchies. By regulating the sexuality of working-class and immigrant women, eugenics would reform the sexual behavior of "women adrift" and limit the procreation of the less "civilized"—that is, nonwhite and working-class—races. And by encouraging middle-class white women to return to full-time motherhood, eugenics would both prevent the new woman from succeeding in her "vain attempts to fill men's places"[36] and ensure that the white race once again would be healthy and prolific.

In order to understand how and why eugenic ideology linked anxieties about race with those of gender, we need to return to the 1915 San Francisco Panama Pacific Exposition, the first popular exhibition in which eugenicists participated. Realizing that a well-attended event such as a world's fair would be an ideal place to popularize the cause of race progress, eugenicists chose the 1915 San Francisco Exposition as the location for the National Conference on Race Betterment. (After the war, eugenicists began to develop displays and organize exhibits at state and county fairs as well.)[37]

Exposition directors were delighted with the choice, believing that the eugenic concern with race betterment appropriately captured the theme of civilized progress promoted by the Exposition. As one director announced to members of the Race Betterment Foundation, "It seems to me that you represent the very spirit, the very ideal of this great Exposition that we have created here." Convinced that the conference would increase attendance at the fair, directors designated one week of the Exposition "Race Betterment Week."[38]

Their predictions were accurate: the Race Betterment exhibit "quickly became one of the fair's most popular exhibits." Located in the Palace of Education, the exhibit provided visible proof, in the form of graphs, charts, photographs, and portraits, of the "rapid increase of race degeneracy." The Race Betterment Foundation, whose membership included prominent educators such as David Starr Jordan and Leland Stanford, expressed delight at the high attendance at its five-day conference. Their sessions, which included topics such as "the commitment of the insane" and "the relation of our public schools to juvenile delinquency," attracted audiences of anywhere from twelve hundred to three thousand. The Foundation estimated that a total of ten thousand persons attended its various programs. In addition, the conference received greater attention from the press than all other exhibits and conferences at the Exposition. The Associated and United Presses generated over one million lines of publicity, and every session received media coverage. The Foundation concluded that "the broad interest on the part of the press certainly indicates a growing interest on the part of the public in the subject of race betterment."[39]

The Race Betterment Conference and Exposition exhibit were highly successful because they tapped into specific concerns of its predominantly white middle-class audience. The exhibit provided vivid visual evidence of the "race-suicide" panic generated by Roosevelt a decade before. But it also reflected two other crises that had emerged since Roosevelt's crusade: first, the onset of World War I, and second, what eugenicists termed the "menace of the feebleminded." Both crises threatened to weaken the race, according to eugenicists. War sacrificed the healthiest, strongest men, while sparing the weak and unfit. The "alarming increase of the insane and mental defectives" suggested that the "feebleminded" were "spreading like cancerous growths . . . infecting the blood of whole communities."[40] Americans began using the term *feebleminded* in the 1850s, when state asylums, or "idiot schools," emerged. Individuals exhibiting a lack of productivity or other behavior deemed by institutional professionals as "backward" were housed in state institutions for the "feebleminded." Yet it was not until the early twentieth century, when eugenics linked feeblemindedness to "race suicide" and the "girl problem," that social commentators expressed anxiety about the "menace of the feebleminded."[41]

The menace, according to many middle-class observers, was a result of human intervention. As Lothrop Stoddard explained in *The Revolt against Civilization* (1923), "In former times the numbers of the feebleminded were kept down by the stern processes of natural selection, but modern society and philanthropy have protected them and thus favored their rapid multiplication."[42] Even Charles Darwin (a cousin of Galton) had expressed concern in *The Descent of Man* (1871) that humans were interfering with natural selection. "If the prudent avoid marriage, whilst the reckless marry," he noted, "the inferior members will tend to supplant the better members of society."[43]

Social observers in the early twentieth century, concerned about what they believed to be the destructive forces on the home front, made the same assumption that Darwin had. The "inferior members of society" were more promiscuous and therefore more prolific than their superiors, they believed, and this promiscuity created the increase of mental and moral defectives. The most visible new figure challenging the manners and morals of Victorian culture—the working-class "woman adrift" or "charity girl"—took much of the blame. Eugenicists believed that the licentiousness exhibited by some working-class women was the result of a mental defect.

These opposing models of womanhood—the mother of tomorrow and the "feebleminded" woman adrift—suggest the importance of gender to eugenic ideology and the desire for race progress. While the mother of tomorrow symbolized the eugenic ideal, the woman adrift represented the dysgenic threat. The eugenic construction of womanhood was double-edged: it contained the potential not only for racial progress but also for racial destruction. Both models carried great symbolic weight in the eugenics movement and would remain pervasive in American culture through the 1950s. The construction and sustenance of these figures point once again to the powerfully ambivalent response to the emerging new woman in the twentieth century: a dynamic figure who challenged the "naturalness" of domesticity and motherhood.

Positive Eugenics: The "Mother of Tomorrow"

In the eugenic vision of racial progress, the mother of tomorrow would save civilization from destruction in two ways. First, she would focus her energies on her domestic duties at home, as wife and mother. By choosing not to abandon domesticity and reproduction even in the face of new opportunities, she would restore the moral forces necessary to keep the American family intact as well as reaffirm male dominance in the public sphere. Second, by choosing procreation she would ensure that the white race would maintain its dominance.

In a fitting finale to the week-long Race Betterment Conference at the Panama Pacific International Exposition, its directors commissioned a theatrical performance that addressed these anxieties about race and gender. Entitled "Redemption: A Masque of Race Betterment," it was described by the Foundation as "A Dramatic Representation of the Great Truths for Which the National Conference on Race Betterment Stands." Performed by two hundred University of California at Berkeley students for an audience of five thousand in an Oakland auditorium, the "morality masque" centered on the struggle of the white race to rise above disease and degeneracy.[44]

As the play opens, Mankind emerges as "conqueror of the world," overly confident in his ability to subdue the forces of nature. Claiming he has achieved complete dominance over the world, he is reminded by Womankind that there "yet is work to be done in the world." She declares that despite his achievements, their position of dominance will not be secure until they overcome "disease, vice and other personal and community ills that make for race deterioration." But Mankind does not heed her call; instead, he chooses to indulge in pleasure and prepares for a dance. Initially opposed to the idea, Womankind is lured by her "love of finery" as Mankind offers her beautiful clothes for the dance, and she relents. Their dance of pleasure grows "more riotous," and they soon forget their duties to the race.[45]

The price they pay for indulging in pleasure is great; their son, Neglected Child, has become crippled by disease "through their own neglect and excesses," and he soon dies. Moaning, Womankind collapses in despair. A chorus of women enters, led by Hope, carrying a beautiful child dressed in white, named Fortunate. Womankind runs forward to grasp the child, confessing her guilt in destroying Neglected Child. Understanding finally that "the salvation of the race lies in the training of youth," she consecrates her life to bringing up Fortunate. As the play ends, Mankind and Womankind "pledge their future to bringing up a race physically perfect and mentally enlightened." They start on a new journey of life, to build up their community based on this eugenic ideal. A chorus of women remains after Mankind and Womankind exit, "chanting of a time of peace and plenty for humankind."[46]

clear gender ideals

Neither the two hundred Berkeley students nor the thousands of spectators could possibly have missed the heavy-handed message of the masque. Two aspects of the Progressive-era American city widely disparaged by social reformers were the commercialization of leisure and the emergence of sexuality for public consumption. The two were not unrelated; as dance halls, amusement parks, and nickel theaters offered unrestricted heterosocial activity for a new generation of youth, the discussion and practice of sexuality moved outside the privacy of marriage and bedroom and into the public sphere. Though social-purity campaigners and Progressive reformers attempted to control public discussion of sexuality and channel it back into marriage and the home, the proliferation of reform literature, as well as the increasing popularity of commercialized leisure and heterosocial activity, served only to widen the public discussion of sexuality.[47]

Yet, as the masque illustrated, extreme dangers resulted from succumbing to such temptation or even merely to representative consumer items (trinkets or fancy clothes, for example) that could lead to such behavior. Once Womankind gives in to her "love of finery," she loses complete control and takes part in the riotous "dance of pleasure." By neglecting her domestic duties and especially motherhood, she loses her child. Devastated, Womankind realizes that she has ignored her racial responsibilities as well. By relinquishing the duties of womanhood for selfish pleasure, she has endangered her race.

Yet Fortunate awaits her: she is given a second chance in the form of another child, and this time she realizes her mistakes. Along with Mankind, she pledges her future to reproductive morality, promising to strive for millennial perfection by bringing up a race "physically perfect and mentally enlightened."[48] Portrayed by a young, healthy, intelligent, middle-class Berkeley student (engaged in the process of "mental enlightenment"), Womankind proved a living and breathing version of the sculptured mother of tomorrow.

Like that sculpture at the entrance of the Exposition, Womankind was portrayed in the masque as the moral force that would redeem the race. Untainted by warfare, she had the strength and virtue to build a unified, healthy family and instill husbands and sons with morality. Moreover, she recognized the significance of her role as mother and did not question her position within the family. The mother of tomorrow reaffirmed the nineteenth-century "cult of true womanhood," which positioned women as arbiters of morality within the home and dissuaded women from asserting too much social and sexual independence.

By the time of the 1915 Exposition, then, womanhood had become a central character in the public discussion of civilization and race progress. Responding to anxieties over race suicide as well as over female sexuality, the eugenics movement suggested that these concerns overlapped. Eugenic ideology appealed to a wide audience because it offered a solution to both assaults on the authority of white middle-class manhood. It reaffirmed racial and gender hierarchies, while at the same time investing white womanhood with a new social responsibility: in Gilman's words, "an active sense of social motherhood." As Gilman explained in the *North American Review* in December 1927, "The business of the female is not only the reproduction but the improvement of the species."[49] Other women, while supportive of feminism and women's rights, recognized the potential danger of abdicating domesticity and motherhood in the name of equality: they would lose their moral authority.[50] "In the general chaos of conflicting feelings [woman] is losing her instinctive adaptation to her biological role as race bearer, and is attempting adaptation to man's reality," physician Beatrice Hinkle commented in 1924. "This is the great problem confronting woman today; how can she gain a relation to both racial and individual obligations, instead of possessing one to the exclusion of the other? Must she lose that which has been and still is her greatest strength and value?"[51]

The mother of tomorrow, an image of womanhood invested with a duty to her race, was a strong and long-lasting symbol of "positive eugenics": the movement to promote the prolific procreation of white middle-class women—those who were considered to be the most mentally and physically sound and who would thus most effectively lead the advancement of civilization. But in order to understand the full extent and appeal of the eugenics movement, we must also look to the target of "negative eugenics": those women whose fertility, eugenicists believed, needed to be restricted. Beginning in the 1910s, this dysgenic threat came to be symbolized by the "high-grade moron," who reformers believed to be "extremely prolific."[52] The construction and promotion of this "high-grade moron" took on almost mythic proportions in the early twentieth century, adding to the appeal and authority of the eugenics movement.

Negative Eugenics: The "Moron"

While the negative-eugenics campaign to restrict the reproduction of the "unfit" was initiated by (predominantly male) scientists, it was bolstered by female Progressive reformers. This new generation of social investigators understood the "problem" of increasingly decadent, unrestrained, working-class female social and sexual behavior as a moral crisis. Social workers, psychiatrists, sociologists, educators, and reformers "ascribed the lax sexual etiquette of the time to the deterioration of women's morals."[53] As young working-class women attended dance halls and cafés unchaperoned, wore makeup, and flirted with men their parents did not know, middle-class reformers looked on with horror.

With the active interventionist mindset of the Progressive era, those concerned about the deterioration of women's morals sought to understand its origins.[54] Were such women "depraved," as early nineteenth-century Protestant clergymen described "fallen women"? Were they merely victims of male sexual lust, as social-purity reformers of the late nineteenth century claimed? By the turn of the century, it was no longer clear how to explain the sexual behavior of women who transgressed genteel standards of civilized morality because civilized morality itself was losing ground.[55]

But the science of eugenics offered a new approach to understanding and treating this "girl problem." Such women were not depraved, sick, or victimized; rather, they were genetically flawed. The eugenic response to the problem of female sexual delinquency was appealing in the Progressive era for two reasons. First, as a science, it appealed to the Progressive belief that science provided "an objective method for resolving social and ethical questions" rather than a moral rationale for regulating female sexual behavior. Second, it suggested a relatively simple solution to the problem. By restricting reproduction of "feebleminded" mothers to eliminate feeblemindedness in future generations, it could curb sexual immorality and advance race progress.[56]

Eugenics gained authority and legitimacy in the early twentieth century, when the rediscovery of Gregor Mendel's laws of segregation and independent assortment led to the establishment of genetics. Working with peas in 1865, Mendel had found that hereditary material is transferred from parent to child. His contemporaries, however, were not impressed, and not until 1900 did scientists appreciate the significance of his findings.[57] Though eugenicists had been arguing for the importance of heredity in their quest for "race betterment" since the 1870s, they had lacked scientific evidence for the transmission of characteristics to offspring. Mendel's laws established genetics as a serious science and lent legitimacy to the eugenic claim that social undesirables—including alcoholics, prostitutes, and even unwed mothers—would produce more of their kind by passing down their supposed genetic flaw to their children.[58]

One American eugenicist in particular made significant use of this claim in his revision of feeblemindedness as both a hereditary disorder and the source of the "girl problem." Henry Goddard, a psychologist and devout believer in the heritability of delinquent and immoral behavior, studied hundreds of cases of "feeblemindedness" at the New Jersey Training School in Vineland beginning in the 1900s. He was troubled by the lack of consensus over diagnosis and treatment of the feebleminded. Without "common criteria," most medical superintendents at institutions for the feebleminded drew on a wide range of approaches to understanding mental deficiency, ranging from physiological to sociological tests.[59] Goddard's interest in locating a standard system for diagnosing and classifying the wide range of mental deficiency he witnessed at the school led him to the psychological studies of French psychologist Alfred Binet. He found Binet's Measuring Scale for Intelligence effective in analyzing his own patients at Vineland and thus proposed that Binet's procedure be used to diagnose the feebleminded at all state institutions. His audience could not have been more receptive: the American Association for the Study of the Feeble-Minded, comprised of institutional physicians and nonmedical personnel, recognized the need for a standard diagnostic procedure and was quick to sponsor Goddard's proposal in 1910.[60]

What Goddard found so effective and original in Binet's method was that it introduced the concept of *mental normality*. Psychologists and physicians had difficulty constructing a coherent cluster of categories to distinguish the wide range of pathological appearance and behavior, Binet argued, because they had nothing standard to measure them against. Without a language or a framework for understanding what normal development was, there could be no hope of understanding the abnormal. So in 1908 the psychologist incorporated such a framework into his testing procedure by "establishing numerical norms for every level of a child's mental growth, based on samples of children's responses. By comparing an individual child's test results with norms established for children of his age, one could determine the child's relative 'mental level.'"[61] It was thus in opposition to normality that mental deficiency came to be understood by physicians and psychologists. In the quest to categorize and diagnose the feebleminded, a new language and a framework for "normal" intellectual development emerged as a focal point.

Developed by psychologists and eugenicists, this language tapped into a burgeoning twentieth-century interest in the concept of normality and the centrality of intelligence to human progress. Though the western frontier of the United States was closing, Progressivism emphasized a new natural resource that required not land but mental cultivation: intelligence. America would continue to expand through the "frontier of science." The key to progress, psychologists declared, was human intelligence. As JoAnne Brown argues, it was not labor or capital that "held real promise as the engine of civilization's advance in the new millennium; the mind, through science, would remake the world."[62]

Within this vision of progress, mental deficiency, in the form of both the waste products of rapid urbanization and the remnants of rural backwardness, needed to be discarded. Physicians emphasized that feeble-mindedness had increased as a result of "industrial and social stress" in cities as well as from a "marked deterioration in the quality of the [rural] population."[63] As normality became a "central organizing principle" of American society, those who displayed abnormal qualities, suggesting an inability to improve, were socially stigmatized as potential threats to the advancement of civilization.[64]

Goddard recognized the ability of this new language of progress and decay to capture public and professional support for the study of mental deficiency. His proposal to use a scale of mental measurement, like Binet's, as a standard diagnostic procedure had widespread appeal in the United States. Between 1908, when Goddard first translated a version of the Binet scale and had it published in America, and 1930, over nine million adults and children were tested using this scale.[65] Standardized mental measurement, best exemplified in the intelligence quotient, or IQ, quickly gained legitimacy in the United States as a scientific procedure and cemented the position of psychology as a serious science. It also provided an essential tool for linking "race suicide" with the "girl problem" because of Goddard's expansion of mental measurement to include the "moron."

Binet's original scale of mental measurement had included two gradations of deficiency: the "idiot," who had a mental age of two or younger, and the "imbecile," who had a mental age of three to seven years. But Goddard was not satisfied that this scale adequately addressed the problem of mental deficiency. He believed the greatest threat to civilization's advance lay with those who demonstrated a mental age of eight to twelve years, a class that had not been given a particular name. This group, consisting of those closest to a "normal" mental age (thirteen or older), posed the greatest danger, in Goddard's opinion.

"We need a name for this high grade group for many reasons," Goddard argued in 1910 in front of the American Association for the Study of the Feeble-Minded. "I presume no one in this audience, certainly none of the superintendents of institutions need to be reminded that the public is entirely ignorant of this particular group." By naming them, Goddard hoped to draw attention to their presence in public school systems throughout the United States, where boards of education were "struggling to make *normal* people out of them" by keeping them in regular classes.[66]

But these boards were making a grave error in treating them as normal, in Goddard's opinion. Even the highest grade of the feebleminded could never become normal, he argued, though they could pass for normal, and thus they were the most likely culprits for spreading the defect to future generations. Rather than trying to disguise or ignore their disabilities, physicians and superintendents needed to underscore them. "One of the most helpful things that we can do," he declared, "would be to *distinctly mark out the limits of this class* and help the general public to understand that they are a special group and require special treatment."[67] In his study of deafness in American culture, Douglas Baynton suggests a similar effort in the Progressive era to single out sign language as deviant. Because of this new emphasis on normality, reformers became increasingly intolerant of the "difference" that deafness or mental deficiency embodied.[68]

Goddard needed a word that would carry scientific legitimacy and arouse public concern, for, as Goddard stressed, physicians needed public assistance in hunting out the high-grade feebleminded. Yet no word in the English language adequately expressed the distinctiveness and urgency of their condition. Goddard therefore constructed his own term from the Greek word for foolish, *moronia,* and the result was the diagnostic label *moron.* "Fool or foolish in the English sense exactly describes this group of children," he announced. "The Century Dictionary defines a fool as one who is deficient in judgment, or sense, etc., which is distinctly the group we are working with [those with a mental age of eight to twelve]."[69] His use of the ancient language gave the term an almost timeless quality, which underscored what Goddard believed to be the permanence of the condition.

Goddard's interest in targeting those he believed to possess a mental age of eight to twelve years old dated back to his graduate training with G. Stanley Hall at Clark University. Hall, like his protégé, sought academic solutions to what he regarded as racial degeneracy and the decline of civilization. In particular, he addressed the "emasculating tendencies of higher civilization": as men evolved into higher beings, they became physically weaker and lost their virility, as evidenced by the emergence of neurasthenia.[70] Hall, like most scientists of the late nineteenth century, believed in the heritability of acquired characteristics (which was overshadowed by Mendelian genetic theory by the early 1900s). As each generation advanced, Hall believed, it would pass its acquired developments on to the next generation, and thus civilization would continually evolve.

Central to this idea was recapitulation theory, which explained how children inherited their parents' acquired traits. As an individual developed, it would "follow the developmental path its forebears took." The more advanced the race, the longer development would take, and the highest stages of human development—advanced intelligence—occurred only within the white races. At the top of the evolutionary scale stood white civilized manhood, with white civilized womanhood just below.[71]

Hall thus proposed that, while developing, young boys should be allowed to embrace primitive savagery as a natural and necessary phase that they would later outgrow. By "encouraging small boys to embrace their primitive passions instead of repressing them," as Hall argued, "educators could 'inoculate' boys with the primitive strength they would need to avoid developing neurasthenia."[72] The most difficult period of development and by far the most dangerous, Hall posited, was adolescence. Linking adolescence to recapitulation theory, he suggested that the growing independence of children eight to twelve years old "corresponded to a lost primitive 'pigmoid' race."[73] Luckily, the more advanced races would outgrow this developmental phase and develop into civilized adults.

Goddard's choice of *moron* to target those with a mental age of eight to twelve years—the age range that corresponded to a more "primitive race" in Hall's theories—was not a coincidence. Goddard came of age after the scientific discrediting of recapitulation theory and what was called "Lamarckism," or "soft heredity" (a theory of biological evolution holding that species evolve by the inheritance of traits acquired or modified through the use or disuse of body parts). Mendelian genetics, or "hard heredity," dismissed Lamarckism and its inclusion of environmental factors, claiming that traits were passed through genes and were entirely independent of the external environment.[74] Goddard thus modernized his mentor's theories by applying them within a eugenic, hereditary framework. A moron was one who could not develop beyond adolescence. He (or she), because of faulty genes resulting in low intelligence, remained trapped in this primitive phase of development.[75]

Thus christened "morons," patients at Goddard's Vineland school and other institutions for the feebleminded across the country were placed in the 1910s in a distinct category of deficiency that posed a challenge to Progressive culture. Yet the term itself was decidedly vague; claiming to include those with a mental age of eight to twelve years, it gave diagnosticians great leeway in determining who fit the category. By adding the "moron" class to the definition of feeblemindedness, Goddard effectively broadened the scope of mental deficiency to include a wider range of symptoms. This new category essentially blurred the distinction between behavior that was unmistakably "normal" and behavior that was "patho-

[handwritten margin note, left:] fear boys would not progress past moron-level development

[handwritten note, bottom:] blurred boundary between normal & pathological - easier to categorize more people as pathological

logical"; it allowed those with new social "symptoms," such as unwed mothers and prostitutes, to be diagnosed as "feebleminded."

Goddard's central evidence for the dangerous and prolific nature of morons was set forth in his own popular work, *The Kallikak Family: A Study in the Heredity of Feeble-Mindedness,* published in 1912. In this study, he traced the ancestry of a young girl (called "Deborah") whom he considered a moron with "immoral tendencies" and found that her genetic flaw could be traced back to her great-great-great-grandmother, a feebleminded tavern girl. From this one tavern girl, he claimed, had come 143 feebleminded descendants, including alcoholics, prostitutes, and criminals. While earlier pedigree studies, such as Richard Dugdale's *The Jukes* (1877), emphasized the importance of environmental influences on human development, the story of the Kallikaks emphasized heredity exclusively. Deborah's great-great-great-grandfather, Martin, had married a prominent Quaker woman after his affair with the tavern girl, and from this union came hundreds of upstanding citizens. Deborah had had the misfortune of coming from Martin's first union and thus inherited the defective gene, while her half-siblings profited from the strong genetic stock of both their parents. Goddard, never one to choose a name without significant meaning, invented "Kallikak" from the Greek words for good, *kallos,* and bad, *kakos,* to emphasize the inevitable destruction of worthy families through a moment of transgression.[76]

This radical redefinition of feeblemindedness, as an outward sign of a fundamental genetic flaw rather than of a slight mental impairment, had enormous implications both for those already housed in institutions for the feebleminded and for those whose attitude, behavior, or appearance would target them for incarceration and sterilization in the future. The person labeled mentally deficient was no longer deemed an object of curiosity or sympathy but a threat to the genetic health and stability of the race. According to this new definition, nothing in the environment—no amount of education, training, or nurturing—could alter the destructive potential stored within a feeble mind. And because "feeblemindedness" had not been a precise diagnostic term to begin with, it was easily transformed into a catchall term for any type of behavior considered inappropriate or threatening. By redefining the boundaries of mental disability to correlate with standards of social and sexual behavior rather than with standardized levels of mental capability, the newly defined term *feebleminded* also defined what constituted normal behavior.[77]

The introduction of the moron category was thus pivotal in the reconstruction of morality as a modern twentieth-century concept replacing nineteenth-century "civilized morality." Since moronia emerged as the binary opposite of normality, and its diagnosis was based as much on moral transgressions as on low intelligence scores, normality became defined as much by moral purity as it did by mental capability. The boundary between what was considered pathological (or backward) and what was considered normal (or modern) was scientifically reformulated on standards of morality.

If moronia was a threat to social order, then morality was the antidote. According to Goddard, what distinguished "us" from "them" was the ability to master morality. Drawing on the work of Hall and on recapitulation theory, Goddard argued that everyone was born with certain instincts that developed "when man was in a more primitive condition. For example, before men came to live together under the conditions of modern society, lying and stealing were virtues, and consequently became instinctive." While Goddard's own society had discarded such values in favor of more "civilized" ones, it could not escape its genealogy. "That [lying and stealing] are instinctive in us to-day no one who is frank and honest with himself for a moment doubts. We have all had the impulses to do these things and probably very few of us have escaped becoming for a time at least victims of these instincts."[78] His message is inclusive: everyone has the potential to fall into this trap. What, then, prevents most from doing so?

Goddard's answer stemmed from Hall's theory of adolescence. For the more civilized, these impulses were merely part of a developmental stage. "Fortunately owing to good training, good environment, we got through that period without disgracing ourselves or the family, and we learned to be moral." Morality was thus an acquired trait, not something God bestowed on certain individuals. "It must be remembered we did *learn* to become moral," he emphasized, "that we overcame these instincts because we were taught and *had minds enough to learn* what was taught us, to see the consequences, and appreciate the importance of our doing as society prescribes."[79] The catch, then, was that moral development came into play only for the normal. The feebleminded were ineligible, by virtue of their supposed limitations. Goddard had modernized recapitulation theory by applying it to mental categories. Civilization would continue to evolve even if acquired characteristics could not be passed down. Only the feebleminded remained trapped in the primitive savagery and immorality of a lost race. As long as the feebleminded were stopped from reproducing themselves, their "race" would die out, and progress would ensue.

In this construction of difference, then, morality is the exclusive privilege of a fully developed mind. As did Hall, Goddard used an evolutionary scale, directly linking the feebleminded with "primitive" ancestors. "Those instincts that lead the child to become what we loosely call a moral imbecile ripen about the age of nine years," Goddard explained. If the mind stops developing at nine years, then "he is a liar, a thief, a sex pervert, or whatever else he may be, because those instincts are strong in him, having already come to full maturity, and the reasoning power, the judgment, those faculties or processes which lead him to learn to control those instincts have never developed and cannot."[80]

Goddard thus presented the problem of race suicide in simple terms that would appeal to his Progressive-era audience. While an earlier generation of middle-class social commentators and reformers merely lamented the growing weakness of manhood and civilization, a new group of scholars, led by Goddard and other eugenicists, addressed the problem of racial degeneration with a new language of social and scientific efficiency and with a new reformist target: womanhood. Intelligence and morality could be quantified. Those who qualified as "normal"— whose mental credentials testified to their moral strength as well—would ensure the advancement of the race. Those targeted as "abnormal"—most often female—would be restricted from reproducing, thus ensuring that their abnormality, now considered to be hereditary, would die out. In 1915 he explained:

It is no undue sentimentalism that assures us that we need to take care of this group of people. We need to study them very seriously and very thoroughly; we need to hunt them out in every possible place and take care of them, and see to it that they do not propagate and make the problem worse, and that those who are alive today do not entail loss of life and property and moral contagion in the community by the things that they do because they are weak-minded.[81]

Women's reform organizations across the country responded to Goddard's call to hunt out the "high-grade feebleminded" and provided much-needed lobbying power. The California Civic League declared: "In view of the fact that feeble-mindedness is the most strongly hereditary thing known and that these moron girls are extremely prolific, the need of custodial care of them is so urgent, indeed so necessary if race preservation is of value, that legislation to that end is imperative."[82] In Massachusetts, the League for Preventive Work published a pamphlet in 1916 entitled *Feeble-minded Adrift: Reasons Why Massachusetts Needs a Third School for the Feeble-minded IMMEDIATELY*. Concerned over what it believed to be twelve thousand feebleminded drifting in and out of schools, hospitals, prisons, and reformatories in the state, the League pressed for a new state institution that would affordably and "effectively remove the immediate social evil and misery" caused by the feebleminded, as well as "absolutely prevent the birth of a new generation."[83] The campaign compellingly emphasized the perceived sexual danger of the female moron. Of the thirty "morons" the League described as plaguing Boston institutions and charities, twenty-eight were women whose uncontrollable sexuality led them to danger and destruction. Some showed a "mania for the company of men," others were "inviting indecent attentions," and some were "living promiscuously, a terrible menace to the men and boys of the town." In every case, proof of their defective mentality came from their social and sexual behavior, which in turn prompted a diagnosis of feeblemindedness.[84]

Eugenicists were delighted at this response from women reformers, for they knew that it would further their cause. As one eugenicist wrote to another in 1913, "If we once aroused the interest [of clubwomen], success would be certain."[85] Taking advantage of higher education and the professionalization of social work, this new generation of women reformers approached urban and rural problems differently from their predecessors. They believed their subjects were no longer victims in need of salvation but clients in need of education, financial relief, and, in some cases, diagnosis and treatment. Social workers replaced child savers, evangelical reformers, and social-purity campaigners in many areas of reform work and, in the process, reconstructed the problem of family violence, unwed motherhood, and juvenile delinquency. Out of this reconstruction came a new understanding of abuse, neglect, immorality, and crime as the consequences of "family pathology." An emphasis on individualism and sexual freedom in urban settings had dire consequences for the American family, moral reformers argued. Inadequate motherhood resulted in defective children.[86]

Thus though not all reformers' subjects were diagnosed as morons, they were scientifically scrutinized in light of their family environment for the first time beginning in the early twentieth century.[87] Whether reformers perceived a subject's problem as biological or environmental, family history took on a greater significance than ever before, particularly a mother's actions or behavior. Concerned with salvaging the healthy, stable American family and the centrality of motherhood, reformers stressed the importance of kinship and bloodlines regardless of whether they perceived an individual's problem to have biological or environmental origins. Either way, they believed that the inadequacies of the working-class home were a cause of female delinquency.[88] Genealogy and motherhood mattered, no matter what the cause of the case.

Beginning in the 1910s, then, feeblemindedness and, in particular, the "moron" category became almost synonymous with the illicit sexual behavior of the woman adrift. Eugenic ideology provided a language and rationale for linking the sexual and reproductive behavior of women with the deterioration of the race. This equation was most forcefully expressed by Dr. Walter Fernald, superintendent of the Massachusetts School for the Feebleminded. In 1918 he declared that "the high-grade [moron] female group is the most dangerous class. They are not capable of becoming desirable or safe members of the community. . . . They are certain to become sexual offenders and to spread venereal disease or to give birth to degenerate children. . . . The segregation of this class should be rapidly extended until all not adequately guarded at home are placed under strict sexual quarantine."[89]

binary

The emergence of two opposing models of womanhood in the 1910s—the "mother of tomorrow" and the "moron girl"—suggests that womanhood itself was undergoing reconstruction in the early twentieth century. New opportunities in education, work, and recreation, as well as evidence that women were controlling conception and limiting family size, allowed for the emergence of a "new woman." Sleek, sexy, and modern, she neglected domestic duties and child rearing and, as a result, generated a concern that women were rejecting the duties of their race. The eugenics movement tapped into this concern, countering the image of the individualistic new woman with the mother of tomorrow, a progressive, forward-looking, socially responsible, moral, and civilized woman who would raise tomorrow's children. Celebrating this image at fairs, such as the Race Betterment exhibit at the 1915 San Francisco Exposition, and during the 1920s with "Fitter Families for Future Firesides" contests, eugenicists familiarized Americans with the notion of "positive eugenics" as a way to counter racial degeneracy. In addition, they promoted the image of the "moron girl" as a biological threat to the health and advancement of the race. Eugenicists argued that her sexual behavior indicated her primitive savagery—trapped in the mind of an adolescent, she was both mentally and morally deficient and a threat to the race.

Eugenic ideology gained popularity and support in the early twentieth century because it linked anxieties about race and gender in the language of Progressivism. As a commissioner of the newly established

American Eugenics Society explained to a colleague in 1922, "Eugenics stands against the forces which work for racial deterioration, and for progressive improvement in vigor, intelligence, and moral fiber of the human race. It represents the highest form of patriotism and humanitarianism, while at the same time it offers immediate advantages to ourselves and to our children."[90] The message was so compelling that even women reformers, themselves blamed for part of the "race-suicide" problem as members of a generation of new women, participated in the eugenic debate. They, too, supported the idea that fertility should be controlled—particularly in the case of "moron girls."

In an era in which people supported increased state intervention and regulation, preventing the procreation of "prolific moron girls" became an accepted response to the problem of female sexuality and the perceived racial decline of American civilization. As chapter 2 will demonstrate, eugenicists fine-tuned their strategies in the 1910s and 1920s on the institutional level, where they had great success in segregating and sterilizing "moron girls," particularly in California.

But though the eugenics movement witnessed more activity and success with "negative" strategies than with "positive" ones in the 1910s and 1920s, it did not lose sight of its goal to influence white middle-class womanhood by promoting motherhood for the benefit of the race. Implicit even in the negative eugenic campaigns was the idea that motherhood should be a celebrated but exclusive privilege—one that had to be earned through a woman's moral reputation.

In 1915, for example, the same year as the San Francisco Exposition, Gilman, residing in California, published *Herland*, a utopian novel in which women live and procreate without the assistance of men. Of central importance in this all-female society is the "New Motherhood," where, as Gilman's biographer Ann Lane points out, "child-rearing is an honored profession permitted only to highly trained specialists."[91] In *Herland*, collective interests transcend the destructive forces of individualism, and all residents prosper in peace and happiness.

This vision of the future reflected a new ideology of motherhood advocated by a wide audience. Many feminists, moral reformers, psychiatrists, cultural critics, and eugenicists perceived motherhood as a privilege, not a right, limited to those who demonstrated the ability of a "highly trained specialist" to inculcate morality, education, and healthy habits in their children. Advocates of what was called scientific motherhood "maintained that maternal instinct needed to be supplemented with scientific education and training." A mother's love was not enough; medical expertise and morality, as measures of ability and genetic health, were essential to child rearing.[92]

The location of the 1915 Panama Pacific International Exposition in San Francisco also suggested the growing importance and influence of California in national and global affairs. The young state sought to promote itself as progressive, and, in the context of eugenics and reform, it was.[93] Its eugenics policies, which were immensely important, are, however, a neglected area of study. Over twenty thousand people were sterilized under California's eugenic-sterilization law, more than in any other state.[94] Many more participated in the debate over sterilization, morality, and motherhood by discussing the issue in school, reading about popularized cases in newspapers, and attending the Exposition. In order to understand how eugenic ideology continued to draw popular and state support long after the Progressive era, we need to turn now to the specific strategies developed by eugenicists on the institutional level.

CHAPTER TWO

From Segregation to Sterilization

Changing Approaches to the Problem of Female Sexuality

While eugenicists promoted the cause of race betterment through popular events such as state and world's fairs, beginning with the 1915 Exposition in San Francisco, they fine-tuned their eugenic strategies in state institutions, away from the public spotlight.[1] Those involved in negative eugenics, the institutional movement to curb the procreation of the eugenically "unfit," recognized that the policies and strategies they practiced on patients would have profound effects outside institutions as well. "It has been recognized for a great many years," the superintendent of the Sonoma State Home for the Feeble-Minded in California declared in 1940, "that state institutions have a far broader scope than just within their own walls."[2] A man whose "indomitable spirit" for eugenics resulted in a record number of sterilizations, F. O. (Fred) Butler was right.[3] The eugenic strategies implemented at the Sonoma State Home—segregation and sterilization—dramatically increased the influence and popularity of negative eugenics in America.

The shifting strategies implemented at Sonoma also reveal the changing response to the "girl problem" between 1890 and 1930. Beginning in the 1910s, eugenicists took part in a cultural debate over the meaning of womanhood in modern America by positing a connection between female sexuality, race suicide, and mental deficiency. Their decision in the 1920s to switch from segregation to sterilization as the primary eugenic strategy reveals a larger shift in the cultural understanding of female morality and social class in the twentieth century. During the Progressive era, eugenicists and other moral reformers advocated the incarceration of sexually promiscuous working-class women. If this small minority of the female population were segregated from society, they theorized, then female sexuality would remain pure, as it had in the late nineteenth century. Quarantining those who challenged the nineteenth-century ideal of female sexual purity would allow civilization to progress.

But by World War I eugenicists and other moral reformers discovered that the "girl problem" had spread into the white middle classes. The turn-of-the-century working-class "woman adrift," who embraced independence and sexual desire, merely foreshadowed a larger shift in the meaning of womanhood that included the legitimation of female sexual desire. Institutional segregation had failed to curb this desire. Eugenicists gave up this battle and instead attempted to channel this newly accepted form of female sexuality into the appropriate places. Indeed, some welcomed sexual modernity and even birth control, as long as they served to stabilize marriage and family and not to weaken them.[4] For the white middle class, sexuality should be limited to marriage and motherhood, they argued. For others, sexuality should be limited to marriage without motherhood. The best way to strengthen morality in modern America, eugenicists believed, was to ensure that both groups met these sexual and reproductive standards. For this reason, they perceived sterilization to be the most effective strategy for curbing the reproduction of the unfit.

This chapter focuses on the history of the Sonoma State Home for the Feebleminded—where, according to one expert in 1942, more eugenic sterilizations were performed on "mental defectives" than in any other institution in the world—in order to illuminate the impact of negative eugenics at the institutional level as well as to explain the movement's growing influence as it moved from segregation to sterilization strategies.[5] The story of the Sonoma Home—its patients, practitioners, methods, and policies—dramatizes the sweeping transformation of mental deficiency from a treatable disease to a sexually loaded, gender-specific, permanent condition requiring either lifelong institutionalization or sexual sterilization.

Founded in 1884, the Home was originally an institution for the education and training of mentally disabled children, "to fit them, as far as possible, for future usefulness."[6] But after the emergence of eugenics in Progressive-era America, Sonoma changed radically. Between 1910 and 1920, the inmate population increased by over 50 percent (making it the fastest-growing public institution in the state), and the largest new "type" targeted for incarceration was the female "high-grade moron."[7] This dramatic institutional growth and the emphasis on female sexuality underscore the conflation of race and gender anxieties in eugenic ideology and explain why female moral offenders would be housed in an institution for the "feebleminded." Henry Goddard's "moron" diagnostic category, which linked mental deficiency with moral deficiency, suggested that immoral behavior was dysgenic and would lead to race suicide. Hence, anxiety about working-class female sexuality was channeled into anxiety about the "menace of the feebleminded."

Though the history of the Sonoma State Home for the Feebleminded is in itself a dramatic and powerful example of the coercive potential of the Progressive-era state institution, its significance extends beyond the history of institutions in America. For eugenicists and moral conservatives in California and throughout the world, Sonoma served as a laboratory where strategies for analyzing and controlling female sexual and reproductive behavior—even outside the institution—were tested. Segregation and sterilization gained worldwide legitimacy as a result of their use at Sonoma. The Human Betterment Foundation (1926–42), a eugenic-research organization that promoted sterilization as a highly effective and successful eugenic procedure in California (discussed in chapter 3), publicized the writings and records of Sonoma's patients, doctors, and social workers in its popular literature and thus helped to convince other states and several European countries to implement eugenic-sterilization laws. Strategies most effective in regulating female sexual and moral behavior—in particular, sterilization—gained legitimacy as Sonoma's patients, parolees, and discharges were touted as scientific evidence of the efficacy of eugenics in negotiating a new, reproductive morality.

The Sonoma State Home

In March 1936, a journalist for the *San Francisco News* visited the Sonoma State Home in order to write a cover story about life inside the institution. The resulting article portrayed Sonoma's patients as dangerous and defective "perpetually child-minded persons" who were sterilized to "prevent the spread of idiocy." According to the superintendent, nearly half the patients were sent there "because they got into trouble in their home communities—delinquencies of sex, stealing, assaults and everything bad that boys and girls can do to upset a town." Most of these troubled cases, he claimed, were "bad girls" sent from the "lowest houses of vice" or after a "series of escapades." The journalist concluded that "a university course in eugenics couldn't match the revelations of a day's visit to California's Sonoma Home." Displayed in front of him in this laboratory setting was scientific evidence that sexual delinquency could be dangerously defective and destructive.[8]

Yet before 1910 Sonoma targeted not "bad girls" but "naturally timid, easily alarmed" children who could not, because of their deficiencies, be properly cared for at home.[9] How did eugenic ideology and the strategies of segregation and sterilization come to take such a central place in institutional policy? More important, how did these changes in policy and practice reflect wider cultural tensions surrounding female sexuality and the fear of racial degeneracy? A close look at the history of the Home reveals the development and influence of eugenics at the institutional level and helps to explain how eugenics became a popular ideal in American society.

Prior to the opening of the Sonoma State Home (originally the California Home for the Care and Training of Feeble-minded Children) in 1885, what are known today as the developmentally disabled were either "confined in jails or almshouses or wandered about aimlessly." But after two San Francisco Bay–area mothers of developmentally disabled children visited New York's School for the Feeble-minded in Syracuse, they set out to establish a similar institution in California. On March 18, 1885, the governor approved "An Act to Establish the California Home for the Care and Training of Feeble-minded Children" to house at state expense "all imbecile and feeble-minded children between the ages of 5 and 18" who had been residents of California for one year and who were "incapable of receiving instruction in common schools."[10]

The state found a permanent site for the Home in 1891: on 1,753 acres of land near Glen Ellen in Sonoma County, where it continues to operate today as the Sonoma Developmental Center. The property, situated at the base of a range of hills, included hundreds of acres of orchards and a vegetable garden, an important source of produce for the institution.[11] Within this rural setting, patients were sheltered from the vice and corruption of urban society. Sonoma's environment was characteristic of late-nineteenth-century social-welfare institutions, which were all in suburban or rural settings for this reason.[12]

The purpose of the institution in the late nineteenth century was to train and educate mentally deficient children and, according to its directors, "to fit them, as far as possible, for future usefulness."[13] Because of widespread concern that the feebleminded were a financial burden on society, superintendents emphasized the importance of education and productivity within the Home. Instead of costing taxpayers money, they suggested, the institution would benefit society by reducing the burden of the feebleminded. Patients were self-sufficient: they grew their own produce, took care of one another, and performed simple tasks on the grounds.

Yet Sonoma's directors also emphasized that the institution was a home, where every child "may be warmly and neatly clothed, tenderly cared for, and ever gently dealt with."[14] Emphasis on a tender, humanitarian setting assured anxious parents that domestic values were central to the institution's operations. Parental roles were stressed as critical to improving the feebleminded, a common emphasis of nineteenth-century asylum superintendents, whose "favorite metaphor was a family one."[15] The Sonoma patients were weak and easily influenced by proper treatment. "Naturally timid, easily alarmed, always so largely dependent upon their instructors for encouragement and assistance, our children are quickly responsive to all our acts of parental care," the directors claimed.[16]

The familial analogy in operation at social-welfare institutions reflected a broader shift in the understanding of family relationships in the nineteenth century. For the first time, childhood was seen as a distinct stage of development. Many parents kept children under their care until their teens or twenties rather than sending them off to work, as had parents in previous generations. In addition, attitudes toward child rearing changed: rather than breaking a child's will, a parent was to shape his or her character. Formation of character was best attained not through punishment, guidebooks explained, but through emotional nurture and by parental example.[17] In the minds of nineteenth-century reformers, social-welfare institutions performed a role similar to that of parents. Responsible for the well-being of patients believed to be in a childlike stage of development, institutional superintendents viewed themselves as parents as well as educators.

The children's apparent "quick response" to treatment at Sonoma reflected the optimism of nineteenth-century institutional reformers, who believed that improvement was always possible. Directors noted that "the incorrigible become tractable, the impudent polite, the rebellious subjective, the dull brightened, the indolent and stupid quickened, until, with our trained and experienced assistants . . . we note an improvement in all classes."[18]

But by the early twentieth century this emphasis on education and care was being replaced by a more custodial policy. Superintendents no longer concerned themselves with the educability or "future usefulness" of a feebleminded patient; instead, they worried about the potential dangers their pa-

tients posed. They shifted their focus from the protection of their patients
to the protection of society. The new science of eugenics and its emphasis
on heredity reshaped popular and professional perceptions of the feebleminded; they were no longer simply a burden but a menace to the gene
pool. As a result, Sonoma adopted a new policy that emphasized the need
to segregate the feebleminded from the "normal" population.

from benefitting to protecting society

Segregation

This gradual shift in policy reflected a new concern about
female sexuality that began in the early twentieth century. The expansion
of the Sonoma State Home reflected this "girl problem" and one possible solution: segregation. California's State Commission in Lunacy, the
State Board of Charities and Corrections, and Sonoma medical superintendent William Dawson targeted women as potential social problems
as early as 1904. Three additional elements contributed to a major esca-
lation in the institutionalization of women as sexual predators over the
next twenty years: first, a growing belief in the medical and scientific communities that feeblemindedness was a hereditary disorder (and thus a eugenic problem); second, the development of mental testing and the use
of the term *moron* at Sonoma to designate a moral transgressor; and,
third, World War I.
    In 1904, Sonoma's medical superintendent urged the state of Califor-
nia to take notice of the problems of the feebleminded and suggested that
these problems were hereditary. "Permit me to say that I do not believe
that mental defectives have received the consideration from the State accorded to the insane and criminal classes," he began, "and yet public safety
demands that these people be housed and cared for to prevent their multiplying their kind, as well as to cut off the source of supply that helps to
fill our jails, reformatories, and insane asylums."[19] By stopping the procreation of the feebleminded, he suggested, the state could virtually eliminate crime and insanity.
    Implicit in his statement was the relatively new argument that feeble-
mindedness was a hereditary disorder. Sociologist James Trent, in his
study of the history of mental retardation in America, locates a dramatic
shift in the medical construction of feeblemindedness in the 1910s. He
argues that the "pitiable, but potentially productive, antebellum idiot and
the burdensome imbecile of the postwar years gave way to the menacing

and increasingly well-known defective of the teens." Trent claims that superintendents began to describe the feebleminded as dangerous because they were afflicted with a hereditary disorder and thus would infect their progeny. The only way to protect society, superintendents argued, was to segregate the defective in institutions.[20]

What Trent misses in his analysis of the "menace of the feebleminded" is its vital link to the "girl problem." It is historically significant that concern with both the feebleminded and working-class female sexuality escalated dramatically in the second decade of the twentieth century. As the history of the Sonoma State Home demonstrates, feeblemindedness became linked with female sexual delinquency in the institutional setting. The California State Commission in Lunacy hinted at the special concern about women when it declared in 1904, "There is no institution, except the [Sonoma State] Home, where it is possible to legally detain a woman who is neither insane nor bad and yet who, by reason of defective mentality, can neither provide for nor protect herself."[21] Believing that women who challenged Victorian standards of morality needed to be segregated, the Commission concluded that the Sonoma State Home offered the best solution.

By 1908, the state of California had heeded Dawson's warning about the importance of the feebleminded. The Commission in Lunacy reported that "this institution has during the last four or five years been receiving the attention its importance demands." Seven new buildings were added to the grounds in response to the "constant demands for admission." Yet the Commission believed more needed to be done to confront the problem of female sexual delinquency. "An effort is being made to locate a cottage upon the grounds of the Home," the Commission explained, "or to create a new institution for the care of girls who are wayward or immoral by reason of defective mentality."[22] The State Board of Charities and Corrections agreed with the Commission's concerns, adding, "There are throughout the State many feeble-minded women of child-bearing age, unmarried, who are giving birth to children who in all probability will be feeble-minded and become State charges."[23] The Board underscored unwed motherhood as a moral offense, suggesting that such behavior itself was evidence of a woman's mental deficiency. Further, the Board shared the widespread assumption that this behavior was hereditary, claiming that children of unwed "feebleminded" mothers would "in all probability" be feebleminded as well.

The State Board of Charities and Corrections, the Commission in Lunacy, and the medical superintendent of the Sonoma State Home expressed two separate concerns about feebleminded women of childbear-

① sex delinquent women → immoral, corrupt communitie
② feeble-minded children → burden state

ing age between 1908 and 1918. First, they believed that these women were engaging in sexual intercourse outside of marriage—an activity that could alone lead to the label of "feebleminded" or "sex delinquent."[24] As a result, the women were "the cause of much immoral corruption in a community."[25] Second, their children would inherit their mothers' defects, thus perpetuating the problem and weakening the race. While the two problems were different—one focusing on the sexual behavior of women, the other on the transmission of supposed mental defects to offspring—a strategy of segregation solved both of them. Indeed, the strategy was both convincing and (for a limited time) effective because it addressed both these problems. While the sexual activity of working-class women was of more immediate concern, the additional threat that their children would carry their defect buttressed the argument for segregation. Eugenics provided the link between the two: it lent scientific credence to the argument for segregating sexually immoral women because it suggested that such behavior endangered the health and strength of the race.

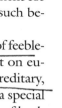

Thus eugenics provided a crucial first step in the construction of feebleminded women as sexual predators. A second step, which built on eugenic ideology and the argument that feeblemindedness was hereditary, came from Goddard's work. His term *moron*, which designated a special class of mental defective, carried in the institution connotations of both immorality and hereditary deficiency. At Sonoma, the moron category, and increased interest in mental classification in general, emerged for the first time in 1914.

In March of that year, the State Board of Charities and Corrections organized a conference to discuss the "Care of the Defectives in the State." As a result of the conference, the Board hired William P. Lucas, professor of pediatrics at the University of California, to conduct a survey of the Sonoma State Home to determine what improvements were needed and to classify patients by their mental ability. Lucas stressed the importance of classification, noting Sonoma's "failure to have the children classified on admission and grouped more according to their potential capacities."[26] Lucas's focus on classification stemmed from a new emphasis in the twentieth century on statistics and measurement and led to an increasing emphasis on "normality" as a central organizing principle of modern civilization.[27]

Along with Professor Lewis Terman, a psychologist at Stanford University, Lucas made a "careful examination of the inmates of the Home with a view to classification of the cases and suggestions for future treatment."[28] Terman played a significant role in the mental testing movement and the eugenic strategy to institutionalize the feebleminded. He was convinced that intelligence was a unitary trait that could readily be measured, and he revised the Binet scale (developed to diagnose the special needs of mentally deficient children in France) for his doctoral dissertation. His final revision, published in 1915 as the Stanford-Binet test, incorporated an index of intelligence that he called an "intelligence quotient," a number he claimed would stay constant throughout an individual's life. This number represented the ratio between the test taker's chronological age and mental age (determined by the test).[29] Terman, a firm believer in the "menace of the feebleminded," claimed that his test would enable psychologists to weed out dangerous defectives. After assisting Lucas diagnose and classify 825 patients at Sonoma in 1914, Terman wrote that "it is clear that society has few tasks more important than that of identifying the feeble-minded and providing their institutional care. There is a growing conviction that society, in self-defense, will be driven to provide institutional care for every feeble-minded individual throughout the reproductive period."[30]

Lucas and Terman noted in their study of Sonoma that the institution had not yet seriously dealt with the "problem of the high-grade moron." After testing 825 patients at Sonoma over four months, they found 281 "idiots," with a mental age of up to two years; 389 "imbeciles," with a mental age of three to seven years; and only 155 "morons." While morons were a minority in the Home, these professionals agreed with eugenicists that they were in fact the most in need of segregation as they posed the greatest threat to society. Terman surmised that "every feeble-minded woman is a potential prostitute," an assumption that led him to stress the importance of diagnosing and segregating female "high-grade defectives" of childbearing age. The key, he argued, lay in the widespread use of his intelligence test:

It is safe to predict that in the near future intelligence tests will bring tens of thousands of these high-grade defectives under the surveillance and protection of society. This will ultimately result in curtailing the reproduction of feeblemindedness and in the elimination of an enormous amount of crime. . . . It is hardly necessary to emphasize that the high-grade cases, of the type now so frequently overlooked are precisely the ones whose guardianship it is most important for the state to assume.[31]

Dr. F. W. Hatch, the general superintendent of California state hospitals, strongly agreed with the findings of Lucas and Terman and used their conclusions to promote the eugenic strategy of segregating sexually suspect women. Sonoma's adoption of Goddard's term, *moron,* to designate the highest grade of mental defect also reflected the belief that moral deficiency and mental deficiency were closely related. In Hatch's discussion of high-grade morons at Sonoma, he noted their "lack of the moral sense" and their inability to "resist the impulse to do wrong." He echoed Goddard's argument that morality was a learned concept and therefore unavailable to the mentally deficient. "The foundation of their troubles," Hatch said of the moron class, "is to be found usually in a true lack of development of brain or mind *engrafted upon them by their ancestors,* which greatly limits their capacities to benefit by study or to properly exercise their will power." By claiming immoral behavior had ancestral origins, Hatch was applying a eugenic argument to justify the segregation of the mentally (or morally) deficient.[32]

Hatch was therefore very much in support of Lucas's conclusion. Based on his survey, the doctor recommended that Sonoma establish a "high grade moron colony for girls," consisting of eight or ten cottages, "each holding 15 or 20 girls, with a housemother and teacher residing with each group." Lucas stressed that the colony should be "erected at some distance from the present group of buildings" to protect other inmates from corruption. Using the medical metaphor of contagion, Lucas contributed to a construction of the female moron as a life-threatening, race-threatening contaminant. Hatch, Dawson, the Board of Charities and Corrections, and the Commission in Lunacy all responded enthusiastically, pressing the state of California to allocate funds for such a colony.[33]

The Board reiterated its recommendation for the establishment of a colony for moron girls in 1916. "The colony," the Board insisted, "would afford [the moron girl] protection, work, and recreation, thus securing for her a happy and profitable life, at the same time relieving society of one of the greatest menaces that now confront it."[34] The colony, or "cottage," plan had become the dominant model for most American institutions beginning in the late nineteenth century; it replaced the earlier model of centralized structures, in which all patients were housed and trained together. The new architectural plan used smaller, separate buildings to distinguish various grades of deficiency and thus illustrated the new emphasis on both specialization and, by the early twentieth century, mental measurement.[35]

The report also hinted that some of these "menaces" would not fall into the moron category when tested. Sonoma had hired a full-time psychologist for the institution, who would address the question "How explain the fact that there are certain high grade morons who test normal but yet are feeble-minded?" In one fell swoop, this question removed the proof of mental deficiency previously required to enter the institution for the mentally deficient. Moral deficiency alone—popularly and professionally accepted as proof of mental deficiency—was evidence enough of the need to institutionalize.[36] Indeed, even as mental testing was gaining authority as a foolproof, practical indicator of feeblemindedness, psychologists were confronted with the problem that not all unwed mothers and prostitutes had IQs in the moron range. Yet, testers believed, their moral depravity was indicative of some sort of deficiency. This paradox inspired mental surveyors in California to devise an alternative scale of intelligence that would rectify the problem.

They received such an opportunity in 1916, when the state embarked on a statewide survey of "mental deviation." The committee noted that "a nation-wide awakening to the menace of the feeble-minded is one of the most noteworthy movements of public thought," and the extent of the problem in the state needed to be determined. Surveyors noted that often the worst offenders had IQs placing them somewhere in the moron, borderline, or even normal ranges. Not satisfied with their results (as it would be difficult to incarcerate large numbers of people who fell into a "normal" category), these authors devised an alternative scale of intelligence that they called "social intelligence." In this category, intelligence was redefined in the "social sense"—"the extent to which the subject is mentally capable of 'managing himself and his affairs with ordinary prudence.'" Offenders, or "persons incapable of doing so, . . . who can not compete in the world . . . on reasonably equal terms," automatically fell into the feeble-minded group.[37]

With this new definition of intelligence, IQ became largely irrelevant to the diagnosis and treatment of social offenders. At the head of this team of surveyors was the man who devoted his career to the development and widespread, standardized use of the IQ test, Terman. Yet he was comfortable with the scientifically questionable approach of a committee frustrated to discover that not every prostitute or unwed mother was a moron. Their solution? "We may ordinarily expect to classify persons as feeble-minded," Terman and his coauthors announced, *whether or not the test results show them to fall within the usual I.Q. limits of that group.*"[38]

Faced with evidence that did not support their assumption that mental and moral depravity were linked, California mental surveyors struggled to redefine the nature of their search. "It can not be too strongly emphasized that the feeble-minded do not constitute a separate and distinct class," Terman declared. "No sharp line of demarcation can be drawn which would separate the feeble-minded from the more intelligent." Yet the surveyors' very objective was to draw such a distinction in order to determine who needed to be institutionalized at Sonoma. Their results suggested the inherent weakness in the use of mental testing to grade morality. Such a study should have undermined the eugenic strategy of segregating moral transgressors through mental testing. Instead, it raised the stakes. If there was no clear boundary between the feebleminded and the normal, then many deviants might be lurking about, unnoticed and difficult to detect.[39]

Investigating a state home for unwed mothers, for example, surveyors found that the mothers demonstrated a "marked inferiority to average adults." In Terman's estimation, these women had become pregnant out of wedlock because they could "pass for normal in almost any community," and consequently "many untrained persons might overlook [their] mental deficiency" and unsuspectingly be tricked into sexual intercourse. Yet one unwed mother had an IQ of 91, placing her in the "dull-normal" range. Terman concluded that while "she will pass for absolutely average-normal in any community . . . she doubtless has weaknesses which intelligence tests do not indicate." Was she to be institutionalized? The surveyors were not sure where or how to draw the line in cases of moral transgressions not backed by mental deficiency. "We are faced with an important problem presented by the small group of *intellectually normal* individuals among these social variants," they admitted.

It is not difficult to understand why a feebleminded girl, such as those described in this study, should become [a victim] of circumstances, and thus be found among the unwed mothers. . . . But that young women whose intelligence is equal or superior to that of ordinary persons of the same age should be found with them and with apparently similar histories, demands that our search for causes shall extend to other fields. . . . weakened will power and excitability seem to have played important parts.[40]

This report, begun in 1916 and published in 1918, reflected the continued unraveling of Victorian notions of feminine virtue during the 1910s. As eugenicists busily targeted the working-class "women adrift" as mentally and morally delinquent in the early twentieth century, they discovered, much to their dismay, that the "problem" of female sexuality—in other words, sexual behavior outside the boundaries of marriage—had spread into the middle classes. By 1918, even "intellectually normal" women, women who "should know better," were becoming unwed mothers, exhibiting a "weakened will power" and "excitability."[41] These surveyors, like many defenders of female social purity in the early twentieth century, did not know what to make of this new evidence of female sexual agency and independence. They were witnessing a transformation in middle-class sexual values that many were reluctant to accept; this transformation made the first two decades of the century a "time of conflict, as defenders of the past and proponents of change contended for hegemony in sexual matters." In New York City during the 1910s and 1920s, for example, the "girl problem" was spreading into the middle class and even the most "respectable" parents found themselves incapable of hindering their teenagers' participation in this sexualized culture or their patronage of heterosocial amusements.[42] Nineteenth-century notions of morality were quickly becoming obsolete.

But if mental testing was proving an inadequate measure of moral deficiency, this inadequacy was easily overlooked in 1918, when the *Surveys in Mental Deviation* report was published. With the U.S. entry into World War I, the female high-grade moron not only was a threat to the family and the race but was now also seen as a threat to national security. In the drive to regulate sexuality and prevent the spread of venereal disease around army camps, information about the role of middle-class women in the transmission of venereal disease heightened public anxiety about the widespread practice of extramarital intercourse. Once again, the Sonoma State Home offered California a temporary solution to the problem of female sexuality.

U.S. involvement in the war unleashed a new series of anxieties about female sexuality. While the eugenics movement and mental testing had helped solidify state and popular support for segregating immoral women, the urgency of war preparations cemented the link between female sexuality and contagion. The *heredity model* of female sexual delinquency (in place at homes for the feebleminded) and the *contagion model* (used by public health officials and reformers fighting venereal disease) merged in the wartime campaign to protect America's soldiers.

Psychologists and superintendents began applying medical metaphors of contagion to stress the necessity of segregating feebleminded women in the 1910s. Equating feeblemindedness with disease and associating immorality with contagion proved an effective scare tactic during the second decade of the twentieth century, when the field of public health had attained professional status. City health departments began to approach the problem of disease control differently because of the new knowledge about and methods of bacteriology. Health policy centered on finding carriers of disease and quarantining contacts, along with the prompt iso-

lation and treatment of cases.[43] In this context, eugenic constructions of
feebleminded women as corrupting society and weakening the human
race with their defective genetic strain triggered a public fear of contam-
ination and garnered increased support for the segregation of the feebleminded. The power of this new segregation was at work at Sonoma,
where administrators began to push for a "moron colony" separate from
the rest of the Home's patients.

[margin note: seg. to protect human race & innocent!]

At Sonoma and other institutions for the feebleminded across the
country, the use of medical metaphors in published reports signaled a
shift in the image of the institution. Homes for the feebleminded appeared not as "institutions" but as "hospitals"; residents were no longer
"inmates" but "patients" or "cases." A patient's entrance requirement, the
intelligence test, became a "thermometer," an instrument to "diagnose"
cases.[44] Walter Fernald, superintendent of the Massachusetts School for
the Feeble-minded and an associate of Goddard, referred to a feebleminded girl as a "plague spot" and called for the segregation of her class
under "strict sexual quarantine."[45] These metaphors also suggested that
the "girl problem" was preventable if the proper measures were taken.

[margin note: conflated with syphilis]

Prior to the war, this conflation of disease, sexuality, and eugenics had
been furthered by the discovery of the syphilis organism in 1910 and the
nationwide movement for the suppression of prostitution. Scientific
proof that syphilis was responsible for debilitating neurological disorders
and was transmissible from mother to child before birth marked it as both
a hereditary and a contagious disease. As such, it effectively bridged the
gap between heredity and contagion models, damaging the race as both
an infectious disease and a genetic disorder.[46] Prince Morrow, a Progressive physician concerned about the "venereal peril," argued that
syphilis "is an actual cause of the degeneration of the race" because of its
effects on offspring.[47] As a degenerative disease, it could turn a person
with a healthy mind into a defective addition to the already overpopu-
lated institutions for the feebleminded. California's General Superintendent Hatch warned in 1918 that "its effects are far-reaching, involving not
only the brain and nervous system, but the body in its entirety." Pointing
out the eugenic factor, he added, "It descends to the children, if not early
and vigorously treated."[48]

Similarities in the portrayal of feeblemindedness and syphilis, both seen
as destructive to the morals of the community and future progeny, paralleled similarities in diagnosis. Just as the intelligence test emerged in the
1910s as the detector for feeblemindedness, the Wassermann test was put
to widespread use in 1911 to detect syphilis. With this new testing device,
the California State Board of Health in 1911 added syphilis to the list of

communicable diseases that warranted treatment and isolation and embarked on a campaign to wipe out venereal disease by eradicating prostitution. From then on, patients at Sonoma were administered both intelligence and Wassermann tests. Though only 4 percent of the population tested positive, the superintendent remained convinced that syphilis and feeblemindedness were related. He explained that he used the Wassermann test "to prove how far syphilis is a factor in these cases."[49]

Beyond all other similarities between feeblemindedness and syphilis, however, the central link between the two was their primary carrier: the female sexual predator. While social-purity reformers of the late nineteenth century sought to protect what they believed to be innocent women from moral ruin by male predators, twentieth-century social workers and eugenicists targeted working-class female sexuality as the source of moral ruin and racial degeneracy.[50] Progressive-era physicians contributed to the construction of the female sexual predator as the source of ruin by associating venereal disease with prostitution. In addition, they often described prostitutes as feebleminded, thus cementing the links between mental deficiency, moral deficiency, and contagion. As one doctor explained, prostitutes would "drain the virility" of men if not stopped.[51] The female sexual predator threatened male virility not only by individual infection but also by weakening the race. The merging of feeblemindedness and prostitution thus heightened concerns about the nation's present and future racial health.

Between 1910 and 1916, vice-commission investigations were conducted in American cities to combat this deadly enemy; these investigations officially closed most of the red-light districts in American cities. One of the largest and most notorious of these was San Francisco's Barbary Coast, finally closed in February 1917 after the enactment of the Red-Light Abatement Law. No longer willing to tolerate commercialized prostitution, the city put eighty-three brothels and over one thousand women out of business.[52]

As war became imminent, reformers and public health officials confronted the problems of prostitution and vice with renewed vigor. In April 1917, the Secretary of War created the Commission on Training Camp Activities (CTCA) to ensure that military training camps were wholesome environments, free from sexual vice and disease. Linking the war at home to the one abroad, one CTCA official remarked that while the "Allies in France have been battering their way forward from one line of defense to another," a "similar campaign is being waged in America" against commercialized prostitution near military camps. The aim of America's war at home, he explained, was to "protect our military forces from prostitutes and other carriers of venereal diseases in order to keep them fit to fight." The strength of the U.S. military depended on the behavior of its civilian population. Thus, any woman who "knowingly tempts a soldier or sailor to immorality" would be considered a traitor to her country.[53]

As with the problem of feeblemindedness, segregation became the key strategy for eliminating prostitution and protecting the soldiers. The Chamberlin-Kahn Bill, enacted by Congress in July 1918, created a civilian quarantine and isolation fund for building institutions for the incarceration of suspected venereal-disease carriers. States used $427,000 in federal funds to construct twenty-seven new institutions and to expand sixteen others. A total of thirty thousand women suspected of illicit sexual activity were apprehended, and over half of these were committed to institutions between 1918 and 1920.[54]

This wartime response to prostitution and illicit female sexual activity brought changes at the Sonoma State Home that underscored the alleged connection between feeblemindedness and prostitution. The Board of Charities and Corrections and the State Commission in Lunacy, along with Sonoma's superintendent, had been requesting funds for a moron girls' colony since 1914, and the additional concerns brought by war lent enough urgency to the situation for their request to be granted. On February 14, 1918, a committee of physicians, charity workers, and county authorities requested funds from Governor William Stephens to expand Sonoma, overburdened with recent admissions.[55] The committee used $15,000, originally intended for a low-grade female cottage, "for the construction of a cottage for the housing of [110] delinquent feeble-minded females from around the army and navy camps as a protection to enlisted men." Sonoma's new medical superintendent, Fred Butler, explained that "this change in the use of the appropriation was made as an urgent war measure, supported by the state government and federal government."[56]

Of 18,000 women incarcerated in the United States during the war, 110 were sent to Sonoma. Why was this Home for the "feebleminded" considered an appropriate location for women targeted as sexually suspect? The *San Francisco Chronicle* noted in its coverage of the deportation of prostitutes to Sonoma that the women had been "an embarrassing charge upon the city and county." According to the chief probation officer of the California juvenile court, the girls were "classified as immoral only because they are weak-minded."[57]

The remarkable overlap in the drives to segregate the feebleminded and to quarantine diseased prostitutes attests to the conflation of disease,

eugenics, and female sexuality in the second decade of the twentieth century. When feeblemindedness specialist Fernald declared in 1918 that high-grade females are "certain to become sex offenders and to spread venereal disease or to give birth to degenerate children," he demonstrated the extent to which assumptions about female sexual behavior, venereal disease, and racial degeneracy overlapped.[58] Were the 110 women sent to Sonoma "immoral only because they were weak-minded," as the probation officer claimed, or weak-minded only because they were immoral? The difference had become largely irrelevant. If feeblemindedness was the cause of prostitution, then the presence of one indicated the existence of the other as well.

Extensive wartime research on the origins of venereal infection, however, unearthed a disconcerting discovery: extramarital sexuality was no longer limited to prostitutes or the "feebleminded." Beginning in 1918, CTCA officials received reports from social workers hired to patrol military training camps about a new source of trouble: "khaki-mad girls." Stricken with "uniformitis," or "khaki fever," these girls flocked to camps for adventure and excitement. This "girl problem" was far greater than reformers or CTCA officials expected, for it had spread into the middle classes.[59]

When used vigorously, both the Wassermann and the intelligence tests revealed an extent of female extramarital sexual behavior that surveyors had difficulty accepting. Just as Terman in his surveys was faced with the problem of "socially inadaptable" but "intellectually normal" girls, wartime antivice crusaders had to confront a new class of offender: young middle-class girls in search of soldiers' companionship.[60] The realization that those forsaking Victorian gender conventions were not only working-class women or prostitutes marked a turning point in the history of sexuality. As more middle-class women refused to abide by Victorian standards of female morality, a strategy of segregation was no longer feasible. Psychologists, eugenicists, and reformers had to come up with a new way to regulate working-class female reproductive behavior in order to prevent further racial degeneration.

Sterilization

As we have seen, the alleged connection between feeblemindedness and prostitution resulted in a widespread wartime campaign to quarantine those believed to be "infected." But the search for these "cul-

· Japanese in WWII
· feeble-minded in WWI
· chinese

prits" helped to undermine such a connection by revealing the extent of female premarital sexual activity in a new generation of women. The shift to sterilization as an alternative to segregation signaled defeat for those opposed to the changes in sexual mores evident by the 1910s. Sexuality was moving out of marriage and family and into the public sphere, where a new ethic of pleasure and self-gratification replaced the nineteenth-century emphasis on sexual self-control.

As sexual standards loosened, many were concerned about what sort of moral system would replace the outdated Victorian morality. "Most people today hold in their minds an image of two worlds—one of gayety and freedom, the other of morality," explained Isabel Leavenworth in a 1924 symposium, *Our Changing Morality*. "It is because gayety and morality are thus divorced that gayety becomes sordidness, morality dreariness. Not until men and women develop together the legitimate interest which both of these worlds satisfy will the present inconsistency and hypocrisy be done away with."[61] How could morality be transformed from its "dreariness"? How could it be modernized?

Eugenicists realized that it was not necessary to eradicate the new ethic of pleasure. Instead, they influenced the making of a modern morality that would curb the impact of this ethic. Individuals could be left free to exercise sexual agency—even the feebleminded woman—as long as the state controlled procreation. Sexually promiscuous women, particularly those of a working-class or immigrant background who would not be able to support children, did not need to be institutionalized for their mental or moral deficiencies as long as they were sterilized. As one advocate proclaimed, sterilization would "ultimately eradicate these people that are undermining our civilization."[62]

But not all eugenicists were initially convinced that sterilization should replace segregation as a eugenic strategy. Believing that it was just as important to curtail promiscuity and venereal disease as to control the procreation of the "unfit," some opposed the use of an operation that might even increase promiscuity and infection. Without the fear of pregnancy, they argued, what was to stop women from increasing their illicit sexual activity?

By emphasizing practicality and preventive medicine and by perseverance, sterilization advocates convinced their more conservative opposition within the movement, as well as much of the public, that sterilization was an effective strategy for advancing the race. Compared with segregation, sterilization was more cost-effective, reached a wider clientele, and did not increase but rather reduced promiscuity, they argued.

In order to understand how sterilization gained legitimacy as an effective eugenic strategy, we need to look again to the Sonoma State Home.

In 1914, even before U.S. involvement in the war had overtaxed Sonoma's patient population, Sonoma was operating at full capacity, with eleven hundred residents and an additional hundred on the waiting list. General Superintendent Hatch noted that "the population of the Home is measured only by its capacity. It never has room to receive all those who apply." He estimated that close to eight thousand feebleminded were residing in California, nearly eight times more than Sonoma could house.[63] Ten years later, Sonoma's population had doubled, and the waiting list stood at a thousand. But a large proportion of those entering the Home, particularly the high-grade female morons, no longer had to wait out their childbearing years within the institution's walls. The practice of eugenic sterilization, firmly institutionalized at Sonoma by the 1920s, dramatically altered Sonoma's segregation policy.

While California enacted the first of a series of eugenic-sterilization laws as early as 1909 (just two years after Indiana passed the first such law in the United States), sterilization was not practiced regularly until 1918, when eugenicists began to see it as the most effective way of combating race degeneracy. Changes in the sterilization law reflect the growing concern over mental deficiency as a threat to scientific and social progress and also the introduction of "normality" as a central and standard principle for measuring this progress. The original act provided for sterilizing inmates of state hospitals and Sonoma, as well as convicts in state prisons, when "such procedure is for the physical, moral, or mental welfare of the inmate."[64] Criticized by the Board of Charities and Corrections as "not broad enough in scope" and without "adequate legal protection," the statute was repealed and replaced with a more effective law in 1913.[65] Hatch announced that under the new law "any inmate of the Sonoma State Home may, upon order of the Lunacy Commission, be asexualized [sterilized] whether with or without the consent of the patient. . . . Thus the way is open, legally, to prevent to a considerable extent the procreation of the unfit. No other one measure means so much to those who are struggling with the problem of the mental defective."[66]

Finally, the law was widened even further in 1917 to apply not only to those "afflicted with hereditary insanity or incurable chronic mania or dementia [but] to all those suffering from perversion or *marked departures from normal mentality* or from disease of a syphilitic nature."[67] As intelligence became the modern natural resource for advancing civilization, "abnormal" mentality, or mental deficiency, suggested backwardness and primitivism. Goddard even convinced other psychologists that the mentally deficient were trapped in a primitive stage of development.[68]

As the state considered all patients at Sonoma to suffer from "marked departures from normal mentality," this amendment allowed Sonoma to implement a widespread sterilization policy under which all residents were potential candidates for the procedure. In addition, the inclusion of syphilis as grounds for sterilization guaranteed that anyone sent to Sonoma—even infected prostitutes who "tested [mentally] normal"—could be sterilized.

Sterilization became law in California largely because of the efforts of one eugenicist, Hatch. As secretary of the State Commission in Lunacy in the 1900s, he drafted a version of the 1909 bill. After it passed, he was promoted to general superintendent of California State Hospitals, where he oversaw implementation of the law. As superintendent, Hatch lost no time in promoting the procedure, ensuring that only sterilization advocates were hired as hospital officials and physicians.[69] "Sterilization may possibly prevent the development of a future genius once in a while," he noted in 1914 after authorizing approximately five hundred of these operations, "but so many who are defective or psychopathic come into the world for lack of sterilization that it is hardly profitable to discuss the question." He concluded that "the longer we continue this work and the more study we give to it, the more convinced we become of its beneficial curative and preventive tendencies."[70]

Hatch oversaw the operations of seven state hospitals: six mental hospitals for the insane and the Sonoma State Home for the Feebleminded.[71] Prior to 1918, only twelve patients were sterilized at Sonoma. From 1902 to 1918, Sonoma's medical superintendent, Dawson, refused to make use of the law. He opposed the measure not on humanitarian grounds but, like many eugenicists and physicians in the 1910s, because he feared the outcome. While he acknowledged that sterilization would "prevent procreation," he also believed it had "a tendency to increase prostitution." Dawson could not condone premarital or extramarital sexual activity by removing the risk of pregnancy, which to him and others served as the last barrier between female sexual morality and sexual decay. Instead, he supported the original strategy of segregation, requesting more "appropriations for buildings to house the feebleminded so that the . . . large number of [applicants could be] admitted."[72]

When Dawson died in 1918, his former assistant, Butler, took over the position; he remained at Sonoma for another twenty-six years. Butler envisioned an entirely different role for the institution, and, once in power, he transformed the Home accordingly. Butler advocated not increased segregation but the widespread use of sterilization. He presented his strategy as not only more progressive but also both financially and eugenically more effective, for a much larger clientele could be reached once the average commitment was a matter not of years but of months.

The transfer of power from Dawson to Butler facilitated the shift from segregation to sterilization. By 1918, many in California found Dawson's policy impractical and outdated. The two eugenically supportive state boards that supervised the state hospitals, the Board of Charities and Corrections and the Commission in Lunacy, welcomed this shift in strategy. "We are very happy to say that Dr. F. O. Butler, superintendent of the Sonoma State Home, is in thorough appreciation of the necessity of sterilization of that portion of feeble-minded inmates who are going forth in the community," the Board of Charities and Corrections announced shortly after his appointment. "Although he has been superintendent but a short time, twenty operations have been performed."[73]

Forty-three years after his promotion, Butler recalled his role as a pioneer in the field. "The first need when I took over in 1918 was water," he commented. "The second was sterilization. I began inside of thirty days."[74] Over the next twenty-six years, he performed one thousand sterilizations himself and supervised a total of fifty-four hundred. Colleagues proclaimed him "a true crusader" and "the most conspicuous physician in the world in this department."[75]

Butler encountered little opposition to his sterilization policy, in part because he presented sterilization as a progressive form of preventive medicine. Noting that the greatest advances in public health had been in this area, he suggested that any physician "who did not believe in preventive measures was way behind the times."[76] Applying this principle to the field of mental hygiene, he proposed that sterilization offered the most effective prophylactic. "It is generally recognized and obvious over the country with people who are at all interested in mental hygiene and the future welfare of our people," Butler wrote, "that sterilization has the best yet to offer in the field of prevention."[77]

Just as the use of medical metaphors of contagion had bolstered support for segregation of the feebleminded during wartime anxiety about venereal disease, the analogy of preventive medicine strengthened Butler's sterilization campaign. In both cases, eugenicists gained authority by framing the problem of racial degeneracy within the larger context of medicine and public health.[78] Butler presented himself not as a surgical zealot but merely as a concerned doctor in search of a cure for the growing problem of degeneracy, evident in the extent of institutional over-crowding. "The over-crowding in our institution and the proverbial long waiting list," Butler recalled, "...urged us to try to do something more than just carry on the routine institution care, training, and treatment of cases without trying to get at the source of the stream that gradually fed cases through the courts to the institution."[79]

In Butler's opinion, sterilization successfully got at that source. The operation "render[ed the inmates] unable to propagate their kind; there-fore, many of them are able to go on parole or be discharged and make their way in the world."[80] In his annual reports to the State Commission in Lunacy, his public addresses, and his published articles, he stressed the efficacy of sterilization on both economic and eugenic grounds. It saved taxpayers' money, he argued, and it prevented procreation. Even years after he had retired, he touted his success at Sonoma, citing the number of patients sterilized.[81] In his hands, sterilization appeared an essential aspect of institutional policy. If anything, he said, "we are not sterilizing, in my opinion, fast enough."[82]

Just seven years after his rise to the superintendency, Butler spoke of the success of sterilization in California institutions in his address to the National Conference of Juvenile Agencies. "We have gradually increased the work and are sterilizing a greater number than ever before," he an-nounced. "The various State departments, the public welfare organiza-tions of the State and the public in general have become interested in this very important subject and have been supporting it, and with this com-bined support I am of the firm belief that sterilization in California has come to stay."[83]

While Butler emphasized the particular importance of sterilizing women, in fact women represented only a slight majority of sterilized pa-tients at Sonoma. During Butler's tenure, between 56 and 62 percent of all sterilizations were performed on women.[84] Patient records reveal, however, both a double standard in Sonoma's sterilization policy and the particular concern about female sexual behavior.[85] Many women were sent to Sonoma because of their perceived "sexual delinquency" and were sterilized for the same reason, while men were commonly sterilized for therapeutic reasons. Men were sterilized for their own benefit, while women were sterilized for the protection of society.

This double standard is most apparent in the discrepancy between who was sent to Sonoma primarily to be sterilized. Of the 249 patients steril-ized between 1922 and 1925, 25 percent of the women and only 2 percent of the men were officially sent only for sterilization. On forty patient records, thirty-eight female and two male, the words "sent for sterilization only" appear under the category "reason for sterilization." This response, of course, was an explanation not for why the patient was sterilized but for why the patient was committed in the first place; it suggested that a large proportion of Sonoma's activities had little or nothing to do with the problem of mental deficiency and much to do with the problem of female sexuality. As Butler testified in 1922, "We receive a great many each year just for sterilization alone, and return them to their respective communities without training, some being discharged, others remaining on parole."[86]

One influential eugenic researcher observed this differential policy at Sonoma and initially publicized it in the May 1927 volume of the *Journal of Social Hygiene*. Dr. Paul Popenoe, a specialist in heredity and eugenics, conducted a survey of sterilization in California for a eugenic-research organization, the Human Betterment Foundation. Writing to the director of the Foundation, E. S. Gosney, Popenoe commented in 1926:

It appears that something like 25% of the girls who have been sterilized were sent up here solely, or primarily, for that purpose. They are kept only a few months—long enough to operate and instill a little discipline in them; and then returned home. It is going to be very interesting to see what becomes of them after that. Some of them are pretty hard characters, and it is difficult to believe that they would follow the straight and narrow path under any circumstances.[87]

Popenoe based his observation on the sexual history of the female patients, a history detailed by physicians and social workers who interpreted the sexual behavior as delinquent. Of 149 female patients, 67, or 45 percent, were classified as sexually delinquent. Their "crime" was often no more than the existence of sexual desire. For this reason, the most common indicator of female sexual deviance found on patient records was the term "passionate," used nineteen times. Other frequently used terms included "immoral" (seventeen times), "promiscuous" (ten times), "masturbator" (eight times), "oversexed," "incontinent," "sly and profane," and "sexually wayward." Only three women were actually accused of sexual crimes, which consisted of adultery, prostitution, and homosexuality.[88]

Doctors took note not only of patients' sexual behavior but also of the sexual organs themselves. For example, of the eighty-two women admitted to Sonoma between January 1918 and August 1919 who were sterilized, forty-one, or 50 percent, were also noted for their "abnormal" genitals. Twenty-two of these patients were singled out specifically for enlarged genitals—the clitoris, vaginal wall, or labia—additional evidence (in the opinion of institutional physicians) of sexual deviance. Indeed, some contemporary doctors, such as Robert Latou Dickinson, believed that female genitals "offered clues for detecting a woman's proclivity toward lesbianism, masturbation, frigidity, and promiscuity." In an influential study of "sex variants," he argued that abnormal genitals were an indicator of a woman's sexual pathology.[89] Enlargement of the genitals suggested not only sexual experience but sexual excess and underscored the assumption that feeble-minded women were indeed "oversexed."

The sexual behavior of male patients, though less closely scrutinized, was also of interest to supervisors. Thirty-seven of the hundred case reports from 1922 to 1925 mentioned sexual tendencies. What is most strikingly different in the male cases, however, is the distinct terminology employed. Not once is a male patient accused of "sex delinquency." Not one was considered "oversexed," "profane," or "sexually wayward." Because sexual aggression (within certain limits) was considered by supervisors to be a normal male characteristic, few were faulted for it at the Home. Sixteen were labeled "masturbators," one an "exposer," three had "immoral tendencies," and ten were labeled as "passive sodomists." Many of those labeled "passive sodomists" were also considered to be "intellectually unstable" or "flighty." Eugenicists were more concerned about passive than active sodomy (as demonstrated by their predominance in Sonoma's records) because passive sodomy connoted a complete reversal of the male sex role, a condition termed "sexual inversion" in the late nineteenth and early twentieth centuries. Terman believed that "passive male homosexuals are typically true inverts in the broadest sense of the term," while the active group seemed not to be. Convinced by sexologist Richard Krafft-Ebing's theory that homosexuals were "stuck at a more primitive stage of evolutionary development," eugenicists targeted this group as weak and "effeminate," as well as a source of degeneration.[90]

not only ♀ = threat

Popenoe observed the difference in the interpretation of sexual misconduct. He noted "the small amount of sexual delinquency attributable to the male feebleminded," comparing his observations at Sonoma with Fernald's at the Massachusetts School for the Feeble-Minded. Fernald found that only seven of his male patients had been "sex offenders—four charged with sodomy, and three with lewdness and the like."[91] Popenoe concluded that patients at Sonoma fell into two groups:

[First,] a group of low grade individuals (predominantly males) who are homeless and helpless and are sent primarily for their own protection; secondly, a group of somewhat higher grade individuals (predominantly females) who have intellect enough to get along in the community but whose behavior is so bad that, after every other course has been tried, they are as a last resort sent to Sonoma primarily for the protection of society.[92]

In studies conducted by the HBF, Popenoe and Gosney emphasized that the higher grade, predominantly female group was a "much greater problem in the community" than the predominantly male low grades. In fact, he argued, the feebleminded male was actually "undersexed" and therefore "unable to compete with more normal, aggressive, and economically competent men for the available women."[93] Their observations underscored the gender differences in Sonoma's population: most men were sent for their own protection; most women, for the protection of society.

Why, then, were males sterilized? If their sexual behavior was not threatening and their lack of sexual aggression indicated a lower procreation rate, why was their behavior considered threatening enough to require sterilization? In order to answer this question, Popenoe and Gosney revealed the double standard in the motive for sterilization. "It is sometimes argued that the sterilization of mentally defective males is without point or purpose," they admitted. "They would not reproduce anyway. . . . This may be true; but since the operation is so simple and harmless, the state authorities have preferred to give the public the benefit of the doubt, by sterilizing."[94] Thus, although both men and women were sterilized, the real concern was the reproductive behavior of women.

The extant patient records and Gosney and Popenoe's analysis bring the patient population at Sonoma into focus. Males and females were frequently there for different reasons. Males targeted as socially disruptive, usually adolescents sent from families who could no longer control them, were usually sterilized, paroled, given job placement, and, if successful, finally discharged. Of 102 males, 24 were sent to Sonoma as a result of petty crimes (including theft and starting fires).[95]

In contrast, only four females had a record of any criminal activity that was not sexual in nature (each was accused of theft). Yet fifty-seven, or 38 percent, were sent because a family member had found them difficult to control at home. For example, one female was considered a "menace" because she "stepped out." Her parents "threw her out when she was pregnant," and, shortly afterward, she was committed to Sonoma. Another began to run after men at age seventeen and "couldn't be kept away from them."[96] The "crime" of women targeted as socially disruptive was a refusal to abide by a standard of sexual morality that was increasingly contested by a new generation of women in the 1910s and 1920s.

Social workers' comments reveal their condemnation of clients' sexual behavior. One noted that her client was "beyond the control of her father and is in need of custodial care." Another observed that her patient had "a history of being hard to control." One female patient was labeled "willful; dishonest; oversexed"; the social worker concluded that she had to be "continually watched." Another was not only "perfectly crazy about boys" but also a "lazy whiner, utterly irresponsible and rotten." Sometimes social workers condemned not only a patient's sexual behavior but also any sign of maternal neglect. One patient not only would "take up with any man she sees" but also "manifest[ed] no love for her children."[97]

Often, social workers' follow-up remarks on paroled patients suggest their class-based criteria for determining whether a patient had earned a formal discharge by renouncing her old ways. One noted the successful case of a former patient:

Annie is now living with her children. She was moved from her old home to this new neighborhood so that she would not be molested by old associates. She has shown a very great improvement in character and even in ability since her return from Sonoma. As near as we have been able to ascertain she has no inclination to revert to her old practices. She has proven that she is a capable mother, and the improvement of the children shows results. She has assured the visitor many times that she is thankful for her short stay at Sonoma.[98]

In this and many other instances, a patient's appropriation of middle-class standards of morality, femininity, and motherhood earned her freedom in the form of an institutional discharge.

The records also suggest the complex power relations involved in the commitment and sterilization procedures. Given the nature and implementation of California's sterilization law, patients had limited access to power and were clearly often victimized. Yet parents also played a role in the dialogue between patient and state; often it was a parent who requested the institutionalization and sterilization of a rebellious teenager. Out of fear or perhaps resignation, parents turned to the state for assistance. In her study of Los Angeles juvenile-court records, Mary Odem finds that parents initiated almost half the girls' cases that came before the court in 1920.[99] While the Sonoma records do not usually reveal how a patient was originally committed, they do suggest the central role of parents. One social worker wrote to Butler in February 1926 regarding the condition of a female parolee. Her comments suggest the extent to which the patient's mother was considered an ally as well as being the final decision maker in regard to her daughter's future.

Marie's mother has found the girl quite difficult. She beats her mother; talks vilely before the other children and refuses to work and obeys no one except her father and as he is away at work most of the time, she is practically a free lance. Mother is interested in having her return to the home but feels it would be well to keep her until after xmas, and we will then make plans to return her.[100]

Parents also played a central role in the decision to sterilize patients; their role reflects the generational conflict over sexual values, as youth in the 1920s (particularly women) challenged the more traditional sexual mores of their parents.[101] Though parental consent was not required by law, Butler made it policy at Sonoma to ask for such consent in order to strengthen the relationship between patients' families and the Home. In 88 percent of the cases between 1922 and 1925, a family member (usually a parent) consented to the operation. In only four cases was there mention of a relative opposing the procedure. One mother "objected vigorously but later consented," another "required much persuasion," while one father was "violently opposed" and another set of parents "objected strongly at first."[102] While it is unclear whether patients were regularly consulted in the matter, Butler claimed that "often a boy or girl would come to ask me if it wasn't time for his or her operation."[103] Regardless of whether Butler's claim accurately portrayed the attitude of patients, the power dynamics among patient, parent, and physician were clearly complex and pose a challenge to the social-control interpretation of institutional history. Patient records suggest the importance of the family in the negotiation of female sexuality at Sonoma as well as the conflict between youth and parents over appropriate sexual behavior.

Generational conflicts between immigrant parents and native-born children proved particularly intense at Sonoma. Though over 90 percent of the patients in 1920 were born in the United States, 50 percent of the males and 39 percent of the females had at least one parent who was an immigrant.[104] One female doctor involved in the treatment of "mentally defective" boys argued in 1930, "We all know that it is the native-born child of foreign-born parents that is apt to be delinquent. The foreign-born parents carry their Old World methods into this country and expect their American children to conform to Old World standards. This conflict of standards (which is further aggravated if one parent is native-born, leading to disagreements between the parents themselves) causes disaster."[105] Thus, both cultural and generational conflicts surrounding female sexuality contributed to a preponderance of first-generation Americans at Sonoma.

While race was a salient issue in early-twentieth-century California, and racism was clearly a central component of the eugenics campaign, patients at Sonoma were predominantly white and native-born. Immigrants were underrepresented in Sonoma's population because of California's extensive deportation program, which returned nonresidents and immigrants to their country or state of legal residence. The California deportation agent noted in 1922 that, "notwithstanding the restrictive immigration laws passed by congress, we continue to receive the average number of mental defectives of Europe." Rather than commit them to Sonoma at state expense, however, the state was able to turn them over to federal authorities. Between 1922 and 1934, an average of thirty-seven immigrants per year were deported under this program; 35 percent of these were Mexican, 13 percent were Filipino, and 8 percent were Italian, findings that point to race as a central factor in the deportation program and that explain why these groups were not targeted at Sonoma.[106]

Unlike the U.S. sterilization policies of the 1960s and 1970s, which were aimed at African Americans, Native Americans, and Puerto Ricans, California's policy did not single out racial minorities or immigrants.[107] While many historians assume that eugenics programs must have targeted nonwhites in the service of race betterment, California eugenics, as evidenced in the history of the Sonoma State Home, focused more on female sexuality and reproduction than on specific racial categories. Yet race itself was of primary importance. Indeed, Progressive-era eugenicists believed that changes in white female sexual and reproductive behavior were primarily responsible for racial degeneration. In institutions such as Sonoma, where the sexual behavior of female patients was a primary focus, the ultimate concern was the preservation of female sexual morality in the white race. Assuming that a nonwhite patient who had violated these standards was merely "exhibiting the usual morals of her race," physicians and social workers would undoubtedly feel little if any anxiety about the implications of her behavior.[108] She was not part of the perceived widespread decline in female sexual morality, because that morality had been an exclusively white female privilege to begin with.

The eugenic strategies employed at Sonoma in the early twentieth century—segregation and sterilization—were responses to changing standards of female sexuality. Initially, eugenicists believed that quarantining the female "high-grade moron" would prevent sexually promiscuous working-class women from infecting the race. Linking moral deficiency to mental deficiency, eugenicists and psychologists used mental testing to prove the primitivism of the promiscuous by suggesting that such women were intellectually trapped in a savage, ancestral mindset. In the years before and during World War I, a medical model of contagion promoted by social hygienists concerned about venereal disease reinforced the image of the promiscuous woman as a sexual predator. Because many associated both "feeblemindedness" and venereal disease with prostitution, the two threats fused into one. Both feeblemindedness and venereal disease would lead to race suicide, and both were caused by changes in female sexual behavior.

But by the 1910s, these changes were no longer limited to the "charity girl"; they had spread into the middle classes. Segregation was no longer an adequate or cost-effective strategy. "It does not follow that all feeble-minded girls should be segregated from society," declared the California Civic League, a forerunner of the League of Women Voters, in 1915. "If they are moral and sterilized there is no reason why they should not be allowed to marry or to remain in their homes."[109] Once the sterilization law had been amended to include all those who demonstrated a "marked departure from normal mentality," the procedure to prevent procreation became an alternative solution that would prevent the spread of feeble-mindedness to future generations.[110]

Yet not all were convinced of the efficacy of sterilization, for it would restrict only procreation, not promiscuity. The practice at Sonoma helped to convince morally conservative opponents that sterilization managed to curb both. While the Civic League argued in 1915 that only those girls who were both moral and sterilized should be released from an institution, doctors and social workers at Sonoma testified in the 1920s that their "sex delinquents" usually stayed out of trouble after sterilization and release. As Superintendent Butler declared, sterilization made them "more amenable to discipline and less restless."[111]

Segregation and sterilization at Sonoma, publicized by the research of Popenoe and the HBF in the 1920s and 1930s, set the stage for a new popular interest in eugenics. It provided proof that motherhood could—and should—be restricted. Like the 1915 San Francisco Exposition, it reinforced the notion that female sexuality and procreation had implications for the entire race and should therefore be carefully monitored. It also provided the statistics and the experience necessary to package eugenic sterilization as a legitimate, indeed necessary, aspect of organized medicine. California eugenicists looked beyond Sonoma's walls in the 1920s and 1930s, eager to gain increased support for eugenics and sterilization from the medical establishment and from the public. How they did so is the subject of the next chapter.

CHAPTER THREE

"Sterilization without Unsexing"

Eugenics and the Politics of Reproduction

By the 1920s, when sterilization was in full force at the Sonoma State Home, a revolution in manners and mores had erupted in America's sexual landscape. The sexual independence and agency exhibited by working-class "charity girls" at the turn of the century had spread to the white middle class. A new generation of women "scoffed at the sexual prudery of its ancestors"—an attitude that underscored a larger shift in sexual meaning: from procreation to pleasure. Sexuality itself was becoming detached from motherhood.[1] As this chapter demonstrates, eugenics played a central role in this shift. In the process of promoting sterilization, eugenicists helped to modernize female sexuality by suggesting that desire, rather than motherhood, was sexuality's primary function. Motherhood was no longer a universal right; it was an exclusive privilege.

In the nineteenth century, prescriptive literature emphasized the importance of sexual self-control in men and "passionlessness" in women.[2] Evangelical Protestants called on middle-class women to act as a source of moral reform; they invoked the term *passionlessness* because they postulated that women were by nature more virtuous than men and possessed less sexual desire.[3] For nineteenth-century males, whom they believed faced greater sexual temptation than did women, the repression of bodily urges would increase moral character.[4] Victorian doctors believed that the body contained a fixed amount of energy. If any organ drew excessively on this limited amount of nervous energy, the body would be depleted, and illness or insanity might occur. Hence, they viewed frequent sexual intercourse and masturbation as a threat to both physical

61

health and moral character.[5] Health reformer Sylvester Graham, in a lecture to young men entitled "Chastity," emphasized the destructive nature of sexual impulse, which concerned nineteenth-century reformers:

Recollect that the *final cause* of your organs of reproduction—the propagation of your species—requires but seldom the exercise of their function! And remember that the higher capabilities of man qualify him for more exalted and exalting pleasures than lie within the precincts of sensual enjoyment!... Who, then, would yield to sensuality, and forego the higher dignity of his nature, and be contented to spend his life, and all his energies in the low satisfactions of a brute when earth and heaven are full of motives for noble and exalting enterprise?[6]

In contrast, sexual theorists of the early twentieth century argued that sexual practice was "neither a threat to moral character nor a drain on vital energies."[7] The central figure in the emergence of this twentieth-century sexual ethos, British sexologist Havelock Ellis (who also considered himself a eugenicist), invested sexuality with new meaning; more than just a reproductive function, sexuality "penetrated the whole person."[8] He believed that as humans advanced from savagery to civilization, their sexual instincts intensified. A truly civilized society, therefore, should not curb desire, as in the nineteenth century, but celebrate it. Gratifying these instincts was "entirely beneficial"; what the world needed was "not more restraint but more passion."[9]

While sexual theorists helped to validate sexual expression as a healthy and legitimate aspect of modern life, middle-class youth put their theories into practice in high school and college, beginning in the 1920s, by engaging in activities such as "petting parties."[10] In addition, Katharine Bement Davis's massive ten-year study, *Factors in the Sex Lives of Twenty-Two Hundred Women,* revealed the extent to which sexuality had moved beyond procreation and "entered the realms of personal desire and intimacy." Of one thousand college-educated women queried, 74 percent admitted to using some form of contraception, despite the fact that information on birth control had "virtually been driven underground."[11]

As a result of this shift in sexual mores, the birthrate continued to decline. Teddy Roosevelt and others had sparked a widespread concern about "race suicide" at the turn of the century, but the white middle-class birthrate dropped even lower in the 1920s. The rate of childlessness reached a record high in the 1920s and 1930s, and the birthrate would not begin to increase until the postwar era.[12] Clearly, positive eugenics, the movement to persuade the fit to procreate, had proved "utterly ineffectual" in the early twentieth century.[13] Eugenicists' claim that women had a reproductive duty to the nation initially could not compete with the sexual culture of the 1920s, which legitimized female sexual desire and separated it from reproduction. As one historian observed, "The middle class had grown soft and selfish, especially middle-class women, who preferred the social whirl to the more serious pleasures of motherhood."[14]

"A Wise Control of Birth Control"

Yet the transformation in attitudes toward sexuality and reproduction was far from complete. While female sexual pleasure was becoming an accepted and commodified aspect of modern identity in the 1920s, birth control was not. The birth-control movement at this time still bore the mark of radicalism, as many believed that the separation of sex from reproduction would undermine public morality.[15] The Comstock law, in effect since 1873, prohibited the importation and mailing of contraceptive information and devices, and full access to birth control would not be available nationwide until the 1960s. Though the declining birthrate demonstrated that many (predominantly in the middle and upper classes) still managed to gain access to birth control, it was not widely talked about in public. Most professionals, including physicians and even feminists, avoided such a controversial topic, which they believed might undermine their credibility.[16] "I regret to say that I cannot give the use of my name in connection with your work," psychologist G. Stanley Hall explained to the Massachusetts Birth Control League in 1916. "If you want to know why, I will tell you frankly that I have borne my share of *odium sexicum....* I have done my bit in this movement and now I am retiring and am going to have a rest from this trouble for the remainder of my life."[17]

Physicians, as James Reed points out, had "no strong motive for a positive attitude toward birth control," often because they, too, were concerned about the declining birthrate and the change in sexual mores. As a result, they "betrayed a startling reticence and lack of information on the subject." They often perceived contraceptives as both morally and physically dangerous. Since its antiabortion campaign in the 1870s, the profession regarded its practitioners as "the gatekeepers of women's virtue."[18]

Even many feminist activists did not believe that promoting women's right to sexual freedom would further their cause. For exam-

[handwritten margin notes:] "Progressivism + public morality"

[handwritten margin notes:] contraceptives to promote immorality and increase fertility rate problems

ple, both the National Woman's Party and the League of Women Voters refused to support birth control because it "clashed with their conceptions of femininity, maternity, and progress." In 1924, the League voted to study sterilization of the unfit as a strategy to reduce degeneracy but refused to include birth control as part of the study because of its controversial nature. Because opponents attacked women's suffrage as a threat to family life, suffragists wanted to avoid any issues that "smacked of promiscuity."[19]

Among eugenicists, there was no consensus on how to approach the issue of birth control. Some were vehemently opposed, pointing out that birth control was responsible for the low birthrate. Others believed that, if used scientifically, it could indeed improve the race. Regardless of their individual positions, the debate generated by eugenicists regarding promiscuity and procreation helped to legitimize female sexual desire and the birth-control movement. For many, this was an unintended consequence of their eugenics campaign. The issue of birth control also further complicates our understanding of eugenics: in this context, it was not a conservative but a modernizing force in reproductive politics, for eugenics lent scientific credibility to the separation of sex from procreation.

Although many eugenicists were opposed to birth control, the ideas and goals of the birth-control and eugenics movements overlapped considerably. Both believed in the importance of sex education and in the overall importance of controlling conception. While eugenicists generally did not want to be associated with the radicalism of Margaret Sanger, most birth controllers supported eugenic goals. Carole McCann's study of the American birth-control movement points out the positive benefits of this association: eugenics provided birth controllers with a scientific language that "helped dissociate birth control from sexual controversy." Placed within a eugenic framework, birth control became a key component of racial progress. In response to the claim made by organized medicine that birth control would lead to a breakdown of sexual morality, Sanger argued the opposite, as she believed that birth control was where "a true eugenic approach to social change must begin."[20]

Out of both a belief in the importance of racial betterment and a desire to gain the support of eugenicists, birth controllers appropriated eugenic language and reasoning in their campaign. Dr. Lydia DeVilbiss, asking for the support of a prominent eugenics organization, wrote the director, "We have found we can get everybody to agree: Every child has the right to be well born." Sanger put the issue even more strongly in eugenic terms, defining birth control at one point as "more children from the fit, less from the unfit." She also helped to ensure that eugenics became a "constant, even dominant theme" at birth-control conferences beginning in the 1920s.[21]

Eugenicists responded with apprehension. To Johns Hopkins geneticist Raymond Pearl, the main question was, "Will the continued practice of birth control have a harmful or a favorable effect upon the race?" Harry Laughlin, superintendent of the Eugenics Record Office in Cold Spring Harbor, New York, acknowledged that birth control would have an impact on the birthrate but feared it was not in the best interest of eugenics. He believed that if it were regulated socially rather than individually, "it would constitute a usable eugenical force of outstanding importance." Social regulators such as "education and family, national and racial loyalties are to be depended upon to convert birth control into an eugenical force of positive value."[22]

This same concern over the prospect of individual regulation led Roswell Johnson, future president of the American Eugenics Society, to become involved in birth-control issues. As he explained to his mentor, Charles Davenport (director of the Eugenics Record Office), "One of the reasons why I have been active in the birth control movement has been to fight individualistic tendencies and to try to help keep this movement as eugenic as possible." At the First National Birth Control Conference in November 1921, Johnson allegedly "forced the issue" by successfully proposing a eugenic resolution. "While desiring a decrease of the world birth rate in general," it read, "this Conference is well aware that this should take place on the part of individuals whose progeny would least contribute to a better race," while "a larger racial contribution from those who are of unusual racial value" should be advocated.[23]

Davenport initially disagreed with his former student, believing birth control to be a "detriment to the best interests of the nation." Despite Sanger's numerous attempts between 1921 and 1929 to convince him to support the cause, Davenport refused, believing birth control to be "committed to a propaganda which is based on insufficient knowledge and concerning which there is grave doubt whether it will work out to the advancement of the race." Like many eugenicists, he feared that the association with Sanger would undermine his own cause.[24]

Clearly, however, birth control was a cause to be reckoned with. Davenport acknowledged that it was "better to face this danger and fight it by instruction . . . rather than to rely upon the futile hope that ignorance will retard the self sterilization of those lines that possess the greatest natural capacity."[25] For this reason, he and many others directed their support not to Sanger but to a man whose interests, they believed, lay closer to theirs: Dr. Robert Latou Dickinson.

Dickinson was a highly respected gynecologist who, beginning in 1882, worked for over fifty years as a practitioner, lecturer, teacher, and researcher. In 1920 he was elected president of the American Gynecological Society, and in his presidential address he criticized practitioners for shirking the albeit "distasteful subject" of contraception and sex instruction. With an "intense interest in female anatomy and sexuality," Dickinson devoted himself to the study of maternal health and female sexuality, and he promoted these issues as founder and director of the Committee on Maternal Health.[26]

As a gynecologist, eugenicist, and birth-control advocate, Dickinson played a central role in bridging the ideological gap among doctors, eugenicists, and birth controllers. Speaking to a group of physicians about birth control in 1916, he remarked, "We as a profession should take hold of this matter and not let it go to the radicals, and not let it receive harm by being pushed in any undignified or improper manner." As historians have noted, he was remarkably successful in achieving this goal, ensuring that birth control would become an issue of family health and stability rather than of female reproductive independence. Distancing birth control from Sanger's radical feminism, Dickinson proposed that doctors take "a wise control of birth control" by using it to promote marital stability and racial health.[27]

To Dickinson, birth control, sexual adjustment, and eugenic sterilization were all part of the same package. "The study of control of conception cannot be dissociated from consideration of sterility, sterilization, and an attempt at definition of the normal in sex life," he argued. As Dickinson, Sanger, and other birth controllers believed, all were part of a broad program of "family regulation in the interests of the parents, the offspring, and the race."[28]

In response to the common charge that the separation of sex from reproduction would undermine public morality, Dickinson, like Sanger, argued that the opposite would occur. Changes in sexual manners and mores had already undermined traditional morality; controlling birth control would result in healthier marriages, better babies, and therefore a healthier race. Regulating conception was therefore socially essential. His work helped to facilitate a shift in the debate about female sexuality and reproduction. By casting reproduction as an issue of "race betterment" rather than of sexuality, he helped put an end to the public reticence about female reproduction.

Promoting Sterilization

Dickinson's role in the birth-control movement is well documented by historians, who note that he was "the mediator through whom organized medicine made its peace with the birth control movement." Surprisingly, however, his emphasis on eugenics has been largely overlooked. Yet it was through eugenics that Dickinson was able to convince medical professionals that controlling conception was not only relevant to organized medicine but also of essential importance. In many ways, Dickinson personified the complexities of the eugenics movement. In certain contexts, he was quite progressive; he was considered both a pioneer of birth control and a "prophet of reason" for his "simple willingness to handle female sex organs and the ability to talk about sex" at a time when organized medicine was "notoriously sex shy."[29] However, he firmly believed that reproduction needed to be regulated by doctors, he actively supported eugenic sterilization, and his techniques were often extremely invasive.

Dickinson's passion for eugenics led him to study female sterilization in California in 1928; he presented his results in an influential paper at the American Medical Association (AMA) later that year. Entitled "Sterilization without Unsexing: A Surgical Review," the paper honored California as the only state to make widespread use of its eugenic-sterilization law, analyzed the various methods used there, and proposed an alternative method of female sterilization. Surrounding him in the conference room were seventeen large (30″ × 40″) charts, maps, graphs, and surgical pictures that illustrated both the simplicity and the necessity of eugenic sterilization.[30] His purpose was to convince the influential AMA to endorse female sterilization as a legitimate, effective, and ethical procedure for preventing procreation selectively in order to improve the race.

Dickinson faced a formidable task. But his successful portrayal of sterilization as a harmless, natural procedure allowed the eugenic-sterilization campaign to take center stage in medical and popular debates about female sexuality and reproduction. By framing female reproductive behavior in the scientific language of eugenics, sterilization advocates such as Dickinson helped to shape the development of modern sexual ideology and reproduction on their own terms.

Dickinson's choice of title, "Sterilization without Unsexing," was significant. In order to convince organized medicine that eugenic sterilization was a legitimate and essential medical responsibility, Dickinson and other sterilization advocates had to present sterilization as a modern procedure that did not interfere with female sexuality. One active eugenicist privately expressed to a colleague the importance of packaging sterilization as a procedure that did not take away sexual desire "because of the widespread public misapprehension that sterilization may 'unsex' the individual." Once it is understood that this is not the case, he continued, "our experience is that the public generally approves of sterilization."[31] Part of the problem was legal terminology: sterilization was initially referred to as "asexualization," and this wording remained in the 1913 and 1917 amendments to California's sterilization law. By the 1920s, however, eugenicists consistently referred to the procedure as "sterilization"—a term that suggested their desire to distance the procedure from any notion of "unsexing."

Sterilization advocates' emphasis that sterilization did not "unsex" in the 1920s reveals the changing relationship of womanhood, sexuality, and reproduction. Was the purpose of sex women's pleasure or procreation? In the late nineteenth century, the standard answer was procreation. Surgical procedures that interfered with female reproduction, such as the "ovariotomy," were attacked by some as violating womanhood by rendering women sterile. As Regina Morantz-Sanchez notes, "Cries of women being 'spayed, unsexed, mutilated,' were common in the 1880s and 1890s, as part of a larger debate over the role of pelvic surgery in women's health." If motherhood was the primary role in society for which women were biologically determined, then the prevention of procreation would destroy their very reason for being. Such procedures "unsexed" women by eliminating the primary function of sexuality as well as the primary function of womanhood. It is important to note, however, that, as with sterilization, those most actively opposed to gynecological surgery were often labeled conservatives, while many female patients and progressive practitioners supported the measure.[32]

But by the early twentieth century, the meaning and purpose of female sexuality were undergoing reconstruction. Sterilization was now presented as a modern procedure that did not interfere with female sexuality because it left a woman's sexual feelings intact. Thus eugenicists argued along with sexologists that pleasure, not procreation, was the primary purpose of sexuality. Cast in this light, sterilization was a progressive measure, which, like birth control, "liberated" women from the biological determinism of the past. Preventing procreation was not a violation of nature, Dickinson believed, but an improvement on it.[33]

However, "unsexing," like sexuality itself, had multiple (and somewhat conflicting) meanings in the early twentieth century; these differences suggest that the debates surrounding the meaning of womanhood and sexuality were far from resolved. The "new woman," who violated the conventional feminine role by attending college and voicing an interest in politics, was often characterized by critics as "unsexed." Indeed, any woman who questioned the centrality of marriage and family in the 1910s risked receiving such a label.[34] Might sterilized women become less feminine and more mannish—a development that would merge the sexes even further? As a derogatory term that underscored widespread anxiety about twentieth-century gender roles, "unsexing" was clearly a dangerous term for the eugenic-sterilization campaign.

Dickinson's push for AMA endorsement of sterilization thus required extreme dexterity. Distancing his proposal from nineteenth-century gynecological surgery as well as twentieth-century slander, Dickinson presented sterilization in an entirely new context. The most common female sterilization procedure, the salpingectomy, involved cutting, crushing, or removing the fallopian tubes; the ovaries remained undisturbed. The operation was intended merely to prevent procreation, not to cure any emotional or psychological disorder, as nineteenth-century procedures had often been.[35] The surgery was relatively harmless, Dickinson argued. "Sterilizing does not involve the removal of any organ or the lessening of sex feeling by the methods now generally employed," he began. "A categorical statement to this effect is necessary, even in a medical article, because of the fixity of the popular belief that mutilating operations are required which result in radical changes in appearance, sensations and behavior."[36]

Yet he believed sterilization also needed to be used more than it was; hence, Dickinson had to convince the AMA to endorse the procedure as legitimate medical practice. Therefore, he justified it on eugenic rather than sexual grounds. Separating reproduction from sexuality, he packaged the sterilization procedure as preventive medicine for race preservation. In order to preserve civilization and democracy, he argued, the most "pressing obligation is to safeguard our biological heritage.... We must have a nation-wide program of selective sterilization for those who are unfit for parenthood."[37] Recast as a eugenic tool to salvage the race by preventing procreation of the "unfit," sterilization became a symbol not of mutilation but of salvation.

And Dickinson became the savior. His quest to "naturalize" sterilization—to make the prevention of procreation more palatable than it was—led him to search for a procedure so simple that patients would think no more of it than of having a cavity filled by a dentist. As early as 1916 he published his proposal for such an operation, which he called "cautery stricture of intra-uterine tubal openings."[38] In this procedure, Dickinson inserted a wire called a "cautery sound" into the vagina, through the cervix, and into the uterus, where he burned the corners of the uterine wall to block entry to the fallopian tubes. Applying the dentist-office analogy, Dickinson positioned himself as the "dentist" of gynecology. "The procedure is shorter than the dentist's clearing of a cavity," he declared in another paper, presented at a conference of the New York Academy of Medicine, "but, like that process, it involves a knowledge of the shape of the rigid walls of a dark cavern and delicate skill of touch." Those most qualified to perform such a procedure, therefore, were gynecologists like himself. "We who date back a half century to routine use of the uterine sound . . . learned to know this cavity."[39]

[handwritten margin note: authority claim]

He developed not only the procedure itself but the instruments required, as well as a test to ensure that the procedure had been successful. "We must invent away complications," he declared to his engineer, who modeled Dickinson's designs.[40] On a trip to Europe in 1926, Dickinson visited "everyone who had made X-ray shadows of the interior of the uterus and the tubes," confirming his prediction that uteri came in all shapes, sizes, and sensitivities. He returned resolved to "run through a series of tests in the dead house [the morgue] that will give us new light on this matter."[41]

Thus when Dickinson was invited in 1928 to participate in an in-depth study of California's eugenic-sterilization program, he eagerly accepted. Awaiting his surgical expertise and eugenic enthusiasm was Ezra Gosney, a wealthy philanthropist, lawyer, and avid eugenicist (figure 1). Gosney was in the process of incorporating a foundation in Pasadena to research and promote the practice of eugenic sterilization; it would be called the Human Betterment Foundation. Gosney wrote Dickinson in January 1926 of his plan to begin such a study, explaining, "We will first have to gather whatever is available as facts in the results of sterilization in California, verify some of them and perhaps do a considerable amount of the same work you wrote me about wishing to do there [in New York]—but do it here, as the first step in our work." Nervous about public response, he wrote, "I am moving cautiously and keeping it all as a confidential matter with the people interested with me as to our ultimate purpose."[42]

Dickinson responded enthusiastically to Gosney's plan and to his invitation to participate in the surgical aspects of sterilization research. "It is good news indeed that you are pushing the investigation of the results of sterilization in California," he wrote. Laughlin, of the Eugenics Record

Figure 1. Ezra S. Gosney (1855–1942), founder of the
Human Betterment Foundation in Pasadena, California.
Courtesy of the Archives, California Institute of
Technology.

Office, a mutual friend, had recently published data indicating that two-
thirds of all sterilizations had been done in California. Dickinson explained that "the proposition I was pushing and asked you to help finance
had to do largely with these cases" in "your enlightened state." Dickinson urged Gosney to draw on Laughlin for assistance.[43]

Laughlin, an influential eugenicist on the East Coast, was encouraged
by Gosney's plan. A few years earlier, he had expressed interest in collab-
orating with Gosney in some "future studies."[44] Gosney had the finances
and the zeal to publicize eugenics in California just as Laughlin had done
in New York. Laughlin therefore advised Gosney carefully. What Gosney
needed to focus on, he stressed, was female sterilization. Laughlin believed that "from the eugenical point of view the sterilization of the
woman is more important than the sterilization of the male." The most

effective way to limit the reproduction of any species, he explained, "depends upon the mass reproductive capacity of the females of the species."[45]

Like Dickinson, Laughlin believed that female sterilization needed to be more accessible than it currently was in order to be widely utilized. The procedure most frequently practiced in the 1920s, a salpingectomy, required opening up the abdomen—a risky, time-consuming measure. He therefore proposed to Gosney and Dickinson that they search for a more effective surgical solution. "One of the most important practical services which could be rendered eugenics just now," he wrote to both of them, "would consist of developing a simplified method for sterilizing the human female." Impressed with Dickinson's gynecological technique and innovative approach to the female body, Laughlin urged Dickinson to take on the project. "The surgical world is looking to you to do this work," he proclaimed.[46]

Dickinson appreciated Laughlin's prodding, for he was eager to perfect and promote his "cautery-stricture" procedure, which he championed on the grounds that it was simple and did not involve invasive surgery. "If we could work out this complete and improved technique of cautery sterilization," he explained, "one can have women and families willing to have the simple procedure done in a large number of cases where an abdominal operation would be refused."[47] With Laughlin's backing, Dickinson hoped to gain the opportunity to expand his research. His response to Laughlin's letter revealed his motivation for a California research trip. Dickinson's research methods were confined to subjects from the morgue, where results were slow in coming. He still did not have the evidence that his cautery-stricture procedure would work effectively on live women. "I wish I might have a chance at 100 women out there in California who could be held or brought back at the fourth or sixth month for tests of tubal patency," he confided, referring to a test he had developed to ensure that the sterilization procedure had effectively sealed off the uterus (figure 2).[48] More to his purposes, he hoped to be able to experiment on human subjects.[49] He continued:

One could also do what has not yet been done. In a few instances of women whose abdomens were to be opened anyway I could, after the abdomen was opened, burn, with my platinum tipped uterine sound, straight through the uterine cornu and determine just what the danger time was. My experiments in the morgue are of course on uteri the subjects of rigor mortis [sic] and without the wet succulence of life, or the easy bleeding from certain mucus membranes in this region.[50]

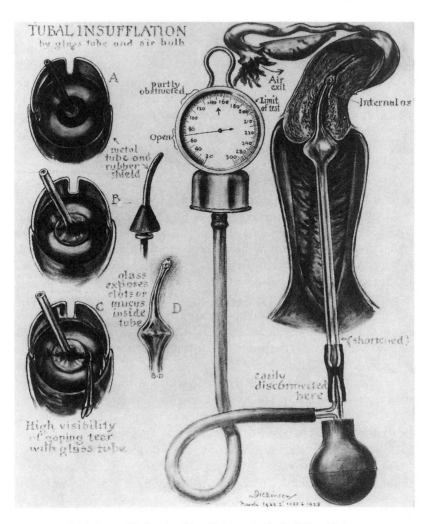

Figure 2. "Tubal Insufflation by Glass Tube and Air Bulb," by Robert Dickinson. Courtesy of the Archives, California Institute of Technology.

Dickinson arrived in Pasadena for a three-week study in February 1928. Gosney and his researcher, Paul Popenoe, had been surveying sterilization in California institutions since 1926. Because Director of California Institutions W. D. Wagner fully supported Gosney's project, he had given Gosney and Popenoe carte blanche to do whatever they deemed necessary with staff, records, and patients at all eight California state hospitals.[51] "This work undertaken by Mr. Gosney will result in great benefit," Wagner explained to the superintendent of each institution, because Gosney was "a man of considerable wealth," interested in getting more states to "carry on the work the same as we are here in California." For this reason, Wagner requested that the superintendents give them "whatever assistance you can."[52]

This courtesy was extended to Dickinson two years later. He visited all eight hospitals, where he interviewed surgeons, observed their surgical methods, and demonstrated his own techniques of cautery stricture and tubal insufflation. Though he believed that cautery stricture would be a more effective sterilization method after extended trials were performed, he was pleased with the results of the more traditional methods in California (figure 3). "I was impressed with the skill and speed of the neurologists who had undertaken the operating," he recalled later.[53] They had sterilized over fifty-eight hundred inmates between 1909 and 1928. Gosney noted additionally Dickinson's enthusiasm "over the large amount of material which we have available here, to throw light on a topic that is of intense interest in almost every civilized country at the present time." Pleased to get such an endorsement from a prominent gynecological surgeon, Gosney promoted his findings immediately. The morning after Dickinson's departure, he informed a local newspaper, "I consider myself fortunate in having gotten the services of a man with such qualifications and prestige as Dr. Dickinson; and I am glad to say that he gives a very favorable report."[54]

Returning to New York, Dickinson surrounded himself with his newfound evidence in the form of interviews, photographs, drawings, papers, and observations (out of which he would later design life-size models of the sterilization procedure); this evidence indicated that sterilization was indeed a successful and cost-effective eugenic strategy. From this experience, as well as his own years of research, Dickinson thoughtfully crafted his "Sterilization without Unsexing" paper for the AMA, in which he promoted the use of sterilization. He argued that based on the results from California, where "more sterilization has been done than in all the rest of the world together," doctors need no longer fear practicing the procedure.[55] And though it was not yet widely practiced, he believed that his own procedure, sterilization by cautery wire, would make sterilization even more appealing to both doctors and patients.

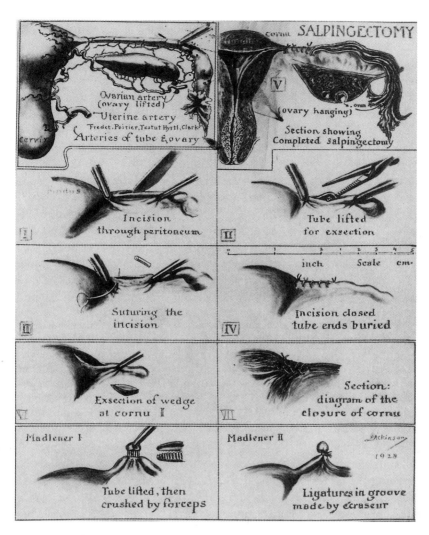

Figure 3. Robert Dickinson's illustrations of various methods of salpingectomy, drawn for his "Sterilization without Unsexing" article, published in the *Journal of the American Medical Association* in February 1929. Courtesy of the Archives, California Institute of Technology.

Sterilization by cautery wire would last no longer and "hurt no more than a dilation of the hot wire to the cervical canal," he testified. "In more than fifty cautery applications I know of no greater disablement than one which necessitated the patient's lying down on a couch for two hours after reaching home." Such language purposely skirted the larger issue of reproductive disablement. Packaging sterilization by cautery wire as a procedure so painless it would be easily forgotten, Dickinson hoped to diminish the impact that permanent loss of reproductive ability might have on prospective patients. "If proven successful as a simple method," he declared, "extended use could be expected, and the main operative objection to sterilizing would be removed."[56]

In addition to simplifying the procedure, Dickinson's method sexualized it as well, underscoring his argument that sterilization did not "unsex." While claiming his procedure was less invasive than abdominal surgery, his description suggests that cautery stricture was actually sexual in nature. The procedure began with the insertion of a cautery sound, "(a thick probe) especially tipped with a tapering spiral of fine platinum wire."

[The probe is] slid up against the external os so as to admit only this length. The possibility of pushing in farther and perforating is prevented by the sliding shoulder fixed where needed. ["As safeguard against too much penetration," he added later.[57]] When the cautery tip has been nestled into the upper angle of the uterus, the current is turned on for from ten to thirty seconds, according to the vascularity of the lining, a succulent uterus needing a longer application of the current. A test is made by starting to draw the sound away. It should adhere rather firmly and bring away a shriveled fragment of tissue all about it.[58]

Burning the uterine cornu sealed the uterus shut, preventing an ovum from entering and thus precluding conception. Dickinson suggested that the deep penetration of the cautery sound did not dry out the sexual organs but heated—and perhaps excited—them.[59] Penetrating both the vagina and the uterus with a "thick probe," exploring the walls of a potentially "succulent" uterus, and doing all of this without anesthesia suggest that Dickinson believed he was actually in some sense "sexing" the patient.[60]

Dickinson's invasive techniques were not limited to sterilization. In his study of "sex-variant" women, Dickinson insisted on close inspection of female genitals and required his assistant to measure genital parts with a ruler and her fingers, as well as to trace the external genitals onto a small glass plate. For a less obtrusive view that did not require a patient's knowledge or consent, Dickinson also kept a hidden camera at the end of his examining table disguised as a flower pot, which he could operate secretly with a foot pedal. He painstakingly produced thousands of detailed sketches of female genitalia and saw himself as an artist as well as a scientist. As Jennifer Terry notes, he signed and dated every drawing, a practice that led one commentator to "call him a cross between Havelock Ellis and Leonardo da Vinci."[61]

It is important to recognize, however, that neither Dickinson's associates nor presumably many of his patients, who faithfully returned to him for years, believed him to be overly invasive or misogynistic in his techniques. He strongly believed that female sexual pleasure was an essential component of modern marriage and continuously promoted sex education. By the 1930s, he even supported self-stimulation with the use of a vibrator for patients suffering from frigidity. Though his use of sexual-response tests and his belief that many women were sexually aroused by pelvic examinations may appear suspect, Rachel Maines's work on the "technology of orgasm" suggests that genital manipulation by a physician in order to induce female orgasm was not unheard of in the early twentieth century.[62]

Because Dickinson was a highly respected gynecologist and sex researcher, his reflections on female sterilization were extremely influential. His search for the perfect procedure—a search he made in order to sell sterilization to the medical community—led him to rethink the relationship among eugenics, sexuality, and medicine. Along with other influential eugenicists who attempted to generate widespread interest in eugenic sterilization in the 1920s, he contributed to a major expansion of eugenic and medical influence over sexuality and reproduction in the twentieth century through his packaging of female sterilization. By pushing for the prevention of procreation on eugenic rather than sexual grounds, Dickinson distanced sterilization from birth control and the stigma of obscenity. Yet at the same time he managed to sexualize the procedure, thus distancing sterilization from nineteenth-century surgeries believed by some to "unsex" and complying with modern notions of sexuality that emphasized the existence and importance of female sexual desire in the context of heterosexual matrimony.

Dickinson's success can be measured by the positive reaction to his paper at the AMA conference. A well-respected practitioner, Dickinson exhibited a professional attitude that "guaranteed a tolerant reception for his unorthodox work."[63] After his presentation, the group voted to organize a thorough investigation of sterilization as a legitimate medical responsibility.[64] Remarkably, this influential body expressed an interest in eugenic sterilization while at the same time condemning birth control and its practitioners for creating an "entirely false sense of values with respect to the important function of childbearing and of parenthood."[65] The main source of medical resistance to birth control stemmed from anxiety that if fertility was too easily controlled, it would reduce the middle-class birthrate even further.[66] Sterilization, however, prevented those believed unfit to reproduce from propagating their kind. Organized medicine supported the separation of sex from procreation predominantly because of its eugenic appeal.

In the 1940s, the AMA continued to welcome papers and exhibits by Dickinson on both birth control and eugenic sterilization. At both the 1942 and 1949 annual meetings, Dickinson was on hand to exhibit his life-sized sculptured teaching models of the "successive steps involved in sterilization operations without unsexing in three-dimensional representation," based on his California research (see figure 4). Eugenicists expressed enthusiasm over the AMA acceptance of Dickinson's models as appropriate for an exhibit. "That such an exhibit can be shown to the public, means a very decided advance over twenty or twenty-five years ago." The AMA continued to support eugenic sterilization until 1960, when the organization's Legal and Socio-Economic Division issued a reappraisal of eugenic-sterilization laws, denouncing the act as a "drastic remedy" as well as a "permanent infringement of bodily integrity."[67]

Though sterilization never achieved the wholehearted support and widespread practice that Dickinson, Gosney, or Popenoe hoped for, it gained considerable attention beginning in the 1920s. Once packaged, promoted, and widely practiced, sterilization gained credence as a legitimate and ethical approach to "safeguard our biological heritage," in Dickinson's words. The research and literature generated by Dickinson as well as Gosney's Human Betterment Foundation (HBF) galvanized public support for eugenic sterilization in the name of race betterment. In 1938, for example, the Foundation (whose board members included Stanford University President David Starr Jordan) received requests from seven thousand college professors, representing every state, for a total of 140,000 copies of *Human Sterilization Today,* their pamphlet on California's eugenic-sterilization program.[68] Teachers requested sterilization literature for a wide range of classes, including zoology, biology, genetics, eugenics, sociology, education, physical education, medicine, parasitology, anatomy, speech, child care, home economics, marital and family relations, and personal hygiene. A professor of psychiatry from the Uni-

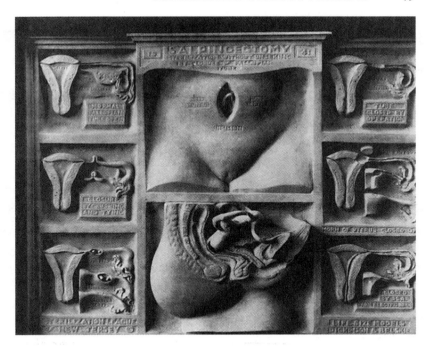

Figure 4. Salpingectomy "sculptured teaching model," designed by Robert Dickinson and sculpted by Abram Belskie and displayed at the American Medical Association convention in June 1942. The model sold for $20, or for $30 with a black carrying case and folding easel included. A vasectomy model was also available. Courtesy of the Archives, California Institute of Technology.

versity of Chicago assigned the Foundation's literature to both medical and social work students, claiming, like many other academics, that the material was "always well received."[69]

Recognizing the need to gain not only medical but also popular support for eugenic sterilization, Gosney and Popenoe published two books on California's sterilization program intended for a lay audience in addition to their regularly requested *Human Sterilization Today* pamphlet. *Sterilization for Human Betterment: A Summary of Results of 6,000 Operations in California, 1909–1929* (1929) and *Twenty-Eight Years of Sterilization in California* (1938) presented the results of Popenoe's surveys of sterilization in California institutions. While Popenoe initially published his results in academic journals, such as the *Journal of Social Hygiene* and the *Journal of the American Medical Association,* he and Gosney wanted to reach a wider audience with their two books. Gosney explained to Dr.

Lewis Terman, an HBF board member, their reason for writing *Sterilization for Human Betterment:*

Our idea in publishing the technical papers was, as you no doubt understand, largely to prepare the way for this popular book. We felt it necessary to have the full evidence on the table where skeptics could see it and weigh it for themselves, before we put out our popular book in which we will necessarily have to make simpler and more dogmatic statements.... The book will reach a much larger circle of readers than do the technical papers, [and] we are particularly anxious that it be as sound, constructive, and immune to attack as possible.[70]

Popular responses suggested that Gosney succeeded in making *Sterilization for Human Betterment* convincing and "immune to attack." New York's Public Health chairman, Jacob Landes, announced on WKBQ in 1930 that "even those of us who feel that sterilization is a rather crude method of dealing with individuals will have to change their opinions on reading this work."[71] Many were not only convinced but also inspired by the book. "It made a profound impression on me," wrote Dr. Signed Dahlstron, "and I am sure it will do it upon every socially interested [person] as well as upon the general public. I feel confident that you are doing a great service toward real human betterment."[72]

Many sterilization supporters congratulated the Foundation on the book's popular appeal. Havelock Ellis, for example, commended Gosney for his success in reaching a popular audience, for "it is the people we need to influence."[73] Letters from college students confirmed that indeed many had been influenced by the eugenic literature (both books and pamphlets) of the HBF. One wrote, "Yours is the first work that I have heard of, in my two years of college, that has struck home with such force.... Everyone that has read your booklet are in accordance with it."[74] Another, studying to be a minister, felt after reading *Human Sterilization Today* that it was "very important for the betterment of the human race as a whole."[75] Thousands of letters from students, professors, ministers, rabbis, social workers, public health and welfare workers, Rotary Club members, physicians, librarians, birth-control advocates, and Parent Teacher Association members expressed personal support and interest in eugenic sterilization as the "only logical and humane method of protecting ourselves" from the rising tide of degeneracy.[76] One prominent eugenicist, Ellsworth Huntington (president of the American Eugenics Society), congratulated Gosney. "You certainly have been most persistent in carrying out your efforts," he wrote. "It is very interesting the way sterilization is being gradually accepted as a necessary measure for preserving the health of the community."[77]

The Effects of Sterilization

Sterilization for Human Betterment appealed to a popular audience because it offered an efficient, scientific solution to the problem of racial degeneracy. By preventing procreation of the mentally and morally deficient, sterilization would safeguard the racial health of communities and ensure that American civilization would continue to progress. "We owe much to California," declared Major Leonard Darwin after reading *Sterilization for Human Betterment*.[78] According to Popenoe, as well as institutional superintendents, physicians, social workers, and patients in California, sterilization appeared to be remarkably efficient in curbing mental and moral deficiency.

How did sterilization gain such support in California? The history of the Sonoma State Home for the Feebleminded suggests some reasons why it became a popular eugenic procedure. Under the leadership of Superintendent Butler, Sonoma staff were able to parole or discharge between one hundred and one thousand female patients per year after they were sterilized. Sterilization was more cost-effective and reached more clients than did segregation.[79]

But sterilization was also initially contested at Sonoma because of the fear that it would heighten female promiscuity. Debates about the effects of sterilization centered on the nature of female sexuality, and they reveal the ambiguities and tensions embedded in the new discourse of female desire. If women possessed sexual feelings, it was "even more dangerous to allow them to have the means to express it without bearing the consequences."[80] Since many women were labeled "feebleminded" because of their sexual behavior, some opponents continued to express concern that sexual promiscuity would increase if there was no fear of pregnancy. *cyclical?*

Thus Dickinson's assertion that sterilization did not "unsex" and in fact had no effect on the patient whatsoever except to prevent procreation was potentially problematic. Though he needed to convince the AMA that the procedure was harmless, this claim would not help persuade those concerned about promiscuity in sterilized women. Supporters of eugenic sterilization were therefore faced with another challenge. It was not enough to provide evidence that sterilization was a safe and effective *why did they care? if these women were no longer a threat, why did it matter if they embodied their degenerate identity?*

procedure. They had to prove that sterilization would bolster, rather than hinder, moral progress. They needed evidence that removing the possibility of pregnancy would not heighten sexual promiscuity. Keenly aware of this problem, Gosney incorporated this research into his survey of sterilization in California.

When Gosney's assistant Popenoe visited each institution and sorted through patient records, he looked for such evidence. "We were particularly anxious to find whether sterilization tends to promote promiscuity, as has sometimes been predicted," he recalled. In order to determine sterilization's effect on the patient, he analyzed the Sonoma State Home records of 605 patients (423 females, 182 males) who had been sterilized and then paroled. With the additional evidence of interviews and correspondence with parole and probation officers, Popenoe categorized the patients as either "successful," "doubtful," or "unsuccessful" on parole. He defined a successful case as one in which "the interests of society were as well protected as if the patient had remained" at Sonoma.[81]

Based on these standards, Popenoe concluded that 65 percent of the female patients and 73 percent of the male patients succeeded on parole. Of those who failed, he determined that there was no proof that sterilization played any part in their inability to succeed outside the institution. Instead, he presented remarkable evidence (which was cited frequently by supporters of eugenic sterilization in sources such as the *Los Angeles Times*) that while nine out of every twelve female patients at Sonoma had been "sexually delinquent" prior to sterilization, only one out of twelve became "sexually delinquent" on parole.[82] In other words, he found that sterilization did not promote female promiscuity but actually helped to reduce it. With this evidence, he could conclude that sterilization, "far from being a hindrance to a mental hygiene program, is a definite help to it; as it is at the same time an unquestioned eugenic gain."[83]

His findings also illustrate the sexual double standard in the selection and treatment of patients at Sonoma. Of the 182 boys paroled, none was considered sexually delinquent. Instead (as noted in the patient records), most were a "nuisance to the community through petty delinquencies and thefts." Most of the 423 girls, however, were "oversexed," had had "frequent sex experiences," were "quite promiscuous," or were "professional prostitutes."[84] Though Popenoe did not specify it, he most likely considered all these attributes to be signs of sexual delinquency.

He found additional evidence in the curative effects of sterilization from doctors and social workers nationwide who wrote both publicly and privately in support of sterilization's curative value. Dr. E. A. Whitney, superintendent of an institution for the feebleminded in Elwyn, Pennsylvania, noted of his sterilized patients that while no marked physiological changes occurred, there were "changes for the better in habits, mentality, and temperament. . . . Most of them 'brighten up' considerably mentally and the majority seem more easily managed and less temperamental."[85] Butler, superintendent of the Sonoma State Home, surveyed his female patients in the 1920s and determined that in 37 percent of the cases sterilization "diminished their sexual desires." He acknowledged that his findings were surprising. "Theoretically, there should be no change, as there is no functional disturbance to the pelvic organs," he explained, "but it appears to make them more amenable to discipline and more easily controlled."[86]

Butler's observation that change occurred when "there should be no change" captured the ambiguities of promoting eugenic sterilization. On the one hand, doctors had to distance the procedure from nineteenth-century gynecological surgery and the charge of mutilation. As a result, they stressed the harmlessness of the procedure. "No organ or gland is removed, no function disturbed, no feeling altered," Popenoe declared in 1928.[87] On the other hand, they had to assure opponents that female patients with "weak inhibitions" would not be encouraged by sterilization to engage in "illegitimate sexual experiences."[88] Therefore, many emphasized physical and mental improvement in their patients who had been sterilized. These contradictory claims underscored the tensions and problems in 1920s sexual behavior and regulation: on the one hand, women possessed sexual feelings (and thus should not be unsexed); on the other, these sexual feelings were potentially dangerous as they could undermine the institutions of marriage and family (thus promiscuity must be curbed). As demonstrated by the birthrate decline, sexuality unleashed from marriage and motherhood in and of itself could lead to what eugenicists termed race suicide.

In order to determine how California doctors and social workers involved in sterilization approached these issues, Gosney and Popenoe distributed questionnaires to all the physicians and social workers who had been connected with the eight California state hospitals since 1909, when the sterilization law went into effect. Fifty-five doctors and twenty-two social workers, nearly all those contacted, responded. Overall, their answers confirmed the result Gosney and Popenoe were looking for: professionals involved in sterilization in California firmly believed that it was a legitimate and useful method for preventing their patients from becoming parents and that it often had curative value as well.[89]

Both physicians and social workers noted physical improvement in many of their patients, even though the procedure — cutting the fallopian tubes — should not have resulted in physiological or mental changes. One doctor shared the assumption with Butler and Whitney that the operations made female patients "more tractable" and believed that it "removes restlessness." Another noted a "general toning up of the system," and nineteen of the fifty-five shared a belief that the operation improved both health and weight (by increasing the appetite). Eight of the twenty-two social workers also believed the operation had physical benefits, noting most frequently "increased physical health and appearance."[90]

Of central concern to Gosney and Popenoe were sterilization's moral effects, and their questions reflected this concern. They asked social workers, "Have you known of any cases where the fact of sterility, with consequent absence of all fear of pregnancy or bastardy, seemed to have acted as an incentive to promiscuity, adultery or other anti-social conduct that would not have occurred had it not been for the sterilization?" All answered in the negative, defending the practice of sterilization as sometimes improving a patient's sexual behavior. One social worker from the Sonoma State Home responded, "Our adult females on parole were sexually irregular, most of them promiscuous and many of them prostitutes prior to commitment to Sonoma. Following asexualization and parole, in no case has a girl under my observation exceeded her former record." Another added, "There are many cases of girls under my supervision who were practicing prostitution before commitment to Sonoma and who are now out earning their way in various occupations."[91]

A majority of the physicians (thirty-one of fifty-five) and eight of the social workers believed that sterilization could produce mental changes in the patient. Social workers noted greater mental stability in some patients, observing that they seemed "more patient, calm, and docile." Most doctors agreed that some patients came out of the operation demonstrating "marked mental improvement," although they acknowledged that "just what the operation in itself has accomplished towards the result is not clear." One doctor believed his patients became "more alert and showed more interest in things and useful employment." Another believed that insane women who possessed a "sex complex are always quieter and easier to manage after being sterilized."[92]

Significantly, Sonoma's one female doctor, Kristine Johnstone, did not agree with her colleagues that the operation produced any mental changes in the patient. She believed that, when asked, many of the women responded positively, each thinking "that it is to her advantage to say her sexual desires have been reduced." As the social workers' reports reveal, a patient was much more likely to be granted parole or discharged if her case was deemed successful by virtue of her supposedly changed sexual behavior.[93] Johnstone's observation suggests the complexity of doctor-patient relations in the institutional setting. Patients did not have a choice about whether they would be sterilized, but they were not merely passive victims. They recognized certain responses were more likely to please doctors and social workers and to result in an earlier release from the institution.

It is also significant that not only male but also female doctors were sterilizing their patients. Based on the names written on the fifty-five doctors' questionnaires, at least eight doctors sterilizing patients in California hospitals were female (at least twenty were male, while twenty-seven signed the form using their first initial only). As Morantz-Sanchez argues, it is overly simplistic to assume that only misogynist male doctors operated on female patients.[94] Eugenic sterilization was often viewed as a legitimate, progressive procedure that was done for the good of the patient as well as of the race.

Indeed, female social workers repeatedly stressed the positive effects of sterilization. One announced that "nothing but good has resulted from the cases under my supervision." Another exclaimed, "I truly feel that every case . . . was to the best advantage of the individual."[95] All twenty-three social workers and all but one of the doctors supported the widespread use of sterilization. "We believe the California law should be strengthened," one social worker wrote, referring to the fact that compulsory sterilization was legal only in state institutions. Another added, "I feel that the California sterilization law is capable of being utilized for much good. It is regrettable that county hospitals do not feel like permitting this operation for their patients as I believe many more needy cases would yield to treatment were it given at such hospitals."[96]

Why did these doctors support sterilization? In his study of psychiatric treatment in California hospitals during the first half of the twentieth century, Joel Braslow argues that most doctors of the insane sterilized their patients for therapeutic, not eugenic, reasons (though his study is limited to Stockton State Hospital). He claims that these doctors gave eugenic reasons for the operation only in order to justify the surgery under the eugenic law. In reality, Braslow argues, "eugenic considerations played only a minor role in the actual clinical decision to sterilize a given individual." But this argument holds true only in the case of male sterilizations, or vasectomies, because of a popular theory in the 1910s and

1920s that the vasectomy would rejuvenate the patient (a theory disproved by the 1930s). Braslow's argument for the therapeutic value of female sterilization is that physicians were concerned about the social consequences of pregnancy in insane women. But this was itself a eugenic concern. Physicians were concerned about the sexual misbehavior of their patients as well as births out of wedlock. Braslow's restricted definition of eugenics (as pertaining directly to hereditary taints) does not allow for an overlap in eugenic and therapeutic rationales for sterilization in California institutions.[97]

Caught up in the belief that a simple surgical procedure could transform Americans into a healthier, more intelligent, and moral race, many doctors, social workers, and others swept up by the eugenics movement and its promise of a scientific solution staunchly defended their right to prevent procreation. "It is my belief that every person of the proper age should be sterilized before leaving an institution of this kind," the medical superintendent of Norwalk State Hospital wrote. He then defended his opinion on medical grounds. "This is in the line of preventive medicine," he continued. "We believe if this procedure were followed in all state hospitals for the insane that in the future it would help to decrease the number of insane patients to a large extent."[98] Another physician wrote, "There are a lot of people coming into the world whom nature never intended to reach maturity, and to prevent this will not only promote happiness but will deliver both individual and society from an unhappy time."[99]

The patients themselves provide a different and crucial angle to the issue. In addition to those patients who, as Johnstone noted, answered questions based on what they believed doctors wanted to hear, others spoke favorably of the operation itself. Some allegedly requested the operation in the first place; such requests suggest that sterilization as a form of birth control had considerable appeal to some patients.[100] Thirty-one former patients responded to Gosney and Popenoe's 1926 request to tell them of their condition (twenty males and eleven females). Seventy percent of the males and all the females who responded noted an improvement in their condition as a result of the operation. "My general health is ten times better," wrote one woman. "I feel fine, I couldn't feel any better." Her response reveals her familiarity with the issue of sterilization for eugenic purposes, as well as some ambivalence about her own position. "I think such operations is just the finest thing there is for people that are not mentally or physically healthy: not only for them, but for all those women who are bearing unwanted and uncared for children," she declared. "One child well cared for is worth more than 10 sick ones." But then she distinguished herself from this group:

Although I do not think my mental sickness had anything to do with that. I fell off a horse once, then a man beat me over the head with a heavy lead of his meat truck, and then that ear operation—all that I know was what caused my head to ache day and night together with my ear which didn't let me sleep and caused me to lose my mind.

Still, she expressed appreciation for the operation.

But my health is ten times better now than before, both mentally and physically. I have been taking piano lessons one month and can play hymns already. My mind is very clear and active. I do all our washing 5 to 6 dozen by hand, and you ought to see me eat when I finish. I rest 10 minutes and I feel as if I could do that much more, and before I couldn't wash three pieces without stopping to hold my back so you can see the difference. My husband is very well satisfied, and we are both living a happy life, which we couldn't before. We have one very health baby 1.5 years old which won first prize on baby contest; he has never been sick that we know of.

She was less satisfied with the permanency of the operation. "We do wish we had a girl," she continued, "but we don't, so we are contented. I think I would have had 12 children at least if it hadn't have been for that operation." Though she supported the theory behind eugenic sterilization, she did not believe she was an appropriate candidate, but she offered personal rather than eugenic reasons for resigning herself to the operation. "I think they all would have been as healthy as the one we have, but I didn't feel we could afford any more [children] on account of my health and because I believe in bringing them up with all care and a good education."[101]

Like most who wrote Gosney to tell of their personal experiences, this woman ultimately approved of the operation not because of its eugenic purposes (though she supported them) but for personal reasons: she felt they could not afford any more children. Another wrote, "My friends all remark how well and strong I am looking. They can hardly believe I am the same person." She also noted that her "mentality [had] improved wonderfully." Stressing the personal benefits of the operation, she recommended it to anyone in a "run down condition." For her, sterilization acted as a form of birth control that she otherwise would probably not have had access to. "I am very happy now not having the fear of giving birth to another baby," she explained, "probably causing me another nervous breakdown like I had when my last child was born." She concluded,

"I am very glad I have gone through with the operation successfully and I think any other woman of like circumstances should not hesitate in having the operation performed: it is only for her own benefit."[102]

Others were more explicit in their appreciation for this permanent form of birth control. "Now that conception is impossible," one wrote, "I have no worry and can enter into relations with my husband with a freedom and zest I never enjoyed before. I believe sexual impulse is stronger since the operation—probably due to the fact that my physical condition is improved in every way." Another added, "As far as sexual life, it is enjoyed with much more pleasure. The operation has been beneficial to me altogether. I feel grateful for it, and I can't thank the doctor and nurse enough for it."[103]

The gratitude expressed by these sterilized women provides a startling contrast to some of the letters published the same year in Sanger's *Birth Control Review*. "Anonymous from Iowa" noted that she was in bad health and lived "in constant dread of having another child." She pleaded to Sanger to tell her "something I can use to prevent this terrible dread that I have with me all the time." "Anonymous from Tennessee" was pregnant for the third time, desperate for some form of birth control; she noted that she would "rather be dead" than give up sexual intercourse.[104]

Though it is difficult to know the exact impact of state and federal anti-obscenity laws on the availability and use of birth control, the thousands of letters that working-class women wrote to Sanger testify to the widespread problem of obtaining reliable contraceptives. Women complained of a lack of information about reproduction and birth control as well as of physicians who refused to tell them about reliable methods and of husbands who abandoned them when they chose incontinence because of fear of pregnancy.[105] For some patients, eugenics may have provided the only way to control fertility.

Scholarship in the history of women's health suggests the need to reevaluate the relationship among doctors, female patients, and surgical intervention. In *Conduct Unbecoming a Woman*, Morantz-Sanchez effectively challenges the historical assumption that female patients were merely victimized by nineteenth-century gynecological surgeons. "Patients' desires and their complex attitudes toward their own bodies did play a part in the overall decision to operate," she argues. Though male doctors clearly did have a great deal of power and authority, they did not always necessarily abuse it. Indeed, it may be easier to understand the gratitude some patients expressed if we "recognize the complexity of this negotiated relationship" between patient and doctor.[106]

In the case of eugenic sterilization, institutionalized patients clearly had less agency than nineteenth-century middle-class women who approached gynecologists for sexual surgery. Yet it is still important to recognize the ways in which meanings behind sterilization and birth control frequently overlapped, particularly for patients. In her study of birth control, abortion, and sterilization in North Carolina, Johanna Schoen notes the "complicated and intertwined history" of these movements, which for too long have been portrayed as coming from opposite ends of the political spectrum. Sterilization, she argues, "was not only a threat to women's reproductive autonomy but also a form of birth control very much desired by women." Atina Grossman's study of sex reform in Germany from 1920 to 1950 also stresses the connections between birth control and sterilization; she points out that during the Weimar Republic "many committed sex reformers and birth controllers came to see sterilization as a positive social good." In Puerto Rico between 1930 and 1960, Laura Briggs notes, feminist activists supported the island's sterilization policy, while Catholics and Nationalists took a conservative pronatalist position against sterilization. During this time period, women received little reproductive health treatment and were apparently pleased with the access to sterilization. Mainland feminists, however, have attacked the program as a form of social control, thereby siding with conservative Nationalists.[107]

Clearly, then, the politics of sterilization are inextricable from the politics of reproduction and birth control. Yet within the American institutional setting, eugenic ideology asserted on both patients and practitioners powerful cultural pressures that decidedly influenced their attitude toward sterilization and, as well, increased their anxiety about the potential for "abnormality." Some patients, for example, may have internalized this ideology and believed that they were unfit to reproduce. The brother of a male patient who had been sterilized wrote to express his support, exclaiming, "I give a sigh of relief when I think there is absolutely no chance for my brother to have any children by any woman." He then claims his brother "says he is very much pleased this operation has been performed [on himself] due to the personal advantages and the undoubted improvement of this practice to the race at large by the weeding out process."[108] Whether these were really the patient's words remains unknown, but they suggest that what the man was told about eugenics and race betterment made him believe that he did not have the right to reproduce.

The testimony of another patient reveals how she had internalized what doctors must have said about her. Sterilized in a California state hospital, she wrote that she was in "better general health" than ever. She was also relieved of the fear "of giving birth to kiddies who would or might not be normal." Though she already had three "bright wholesome youngsters," she did not want to "chance it again." She added that she was "very much in favor of this method of preventing propagation of those who are unfit mentally, morally, or physically." Though she did not identify which category she fell into, presumably she viewed herself as being unfit to reproduce.[109]

Concern that potential children "might not be normal" was not limited to institutional patients. The eugenic quest for human mental, moral, physical, and racial perfection contributed to a growing anxiety about individual potential as well as a developing interest in normality as an organizing principle. Mental testing, initially designed to weed out "undesirables," developed into a formal educational tool to measure, rank, and standardize intelligence. The California sterilization law contributed to this normalization process by singling out anyone exhibiting "marked departures from normal mentality."[110]

By arguing that only select bodies were fit to reproduce, eugenicists added unprecedented pressure toward cultural uniformity and contributed to the increasing stigmatization of difference. Restricting motherhood to those who fit uniform standards of mental, moral, and physical fitness, eugenicists succeeded in inculcating a new "reproductive morality"; this standard convinced some Americans who did not meet it that they should not reproduce.

The letters of patients and prospective patients testify to the eugenic success in inculcating this new morality. Eugenicists, concerned about racial degeneracy, a declining birthrate, and morality in a culture undergoing a revolution in sexual meaning and in gender roles, accepted some of these changes. They acknowledged the recent argument that female sexual desire was a normal and natural component of female desire. They also accepted, indeed promoted, the separation of sexuality from reproduction. However, they declared that reproduction was an issue of racial health, not of individual choice. While sexuality was a matter of personal freedom, reproduction was a collective issue whose goal, they suggested, was the betterment of the human race.

Thus some patients who were sterilized may not have voiced an objection because they had internalized this reproductive morality and its argument that individual reproductive rights should be subordinate to the collective racial health of the nation. Evidence of this internalization extends beyond patients to those who feared that they might have a hereditary deficiency and so thought that perhaps they should be sterilized or should choose a different mate. These individuals, or sometimes their prospective mates, wrote the HBF in the 1930s and early 1940s (before the Foundation closed after Gosney's death in 1942) in search of information on sterilization regulation and procedure.

For example, one young student wrote the Foundation in April 1940 after having read Gosney's pamphlet *Human Sterilization Today* (widely distributed at colleges and universities as a teaching tool) and on the advice of her professor. "I am engaged to a young man who is very brilliant mentally and in very good health physically," she began and then continued:

However, he does have one physical defect. His ears are smaller than the normal size and have somewhat the appearance of being upside down. This defect as far as I know has not been hereditary. I have been advised by a physician that if we have children it may result in something degenerate. I have consulted many books and talked with my professors and they all agree that a child may have the same type ear but would not result in anything worse than that. I would certainly appreciate any advice you could give me. I haven't mentioned this to the boy because as the Doctor told me, it may not result in anything and would only cause worry on his part. I'm very anxious to have advice on this problem, so I would appreciate hearing from you about this.

Gosney responded supportively, believing that she would be safe under the circumstances.[111] Her anxiety, heightened by the physician and her knowledge of eugenics, influenced a major decision: whether to marry and have children with a man whose slight physical "defect" could spread to their children.

This evidence suggests that determining whether those sterilized under eugenic laws were victims or active agents in search of permanent birth control for their own benefit may be overly simplistic. Eugenic ideology, preached in universities, advocated by physicians and surgeons, and promoted by organizations such as the HBF, emerged as a pervasive force by the 1920s, influencing individual decisions about marriage and family and altering perceptions about the relationship between personal identity and the human race. It became important to act and appear "normal," to be measured and evaluated in relation to a national standard; applying this standard ultimately resulted in a "very narrowly defined profile that describes only a minority of actual people."[112] The additional pressure to produce "normal" children generated enough anxiety in some to make them question whether they should ever become parents.

relationship between personal
identity & human race shifts

For example, Stanford Vieth, living in Hollywood, California, wrote Gosney at the HBF to express concern about having children. "My wife and I are both extremely tall," he began, "and this also worries us as we do not wish to bring abnormally tall children into the world. My wife is six feet and three-quarters of an inch tall and I am six feet and four inches tall." Recognizing that this was a hereditary trait, he continued, "We have traced our height through both of our families and find that all of our relatives have been very tall. All of which leads us to believe that our offspring would naturally be tall."[113] His concern rested on the belief that to be abnormally tall would be reason enough not to exist in the first place, a powerful example of the extent to which eugenics and its emphasis on reproductive morality, as well as normality, pervaded the private thoughts and decisions of Americans beginning in the 1920s.

Another letter, from Miss M. Walsh in Santa Barbara, revealed the concerns of a young woman in search of companionship and willing to be sterilized for it. "I understand that you would be able to advise me about sterilization," she began. "I am a young woman, 30 years old, and I am totally deaf. Now I should like to marry but I feel that because of my affliction children would not be desirable. Yet I am sure that I should have as much right to the companionship of marriage—as any other person."[114] Walsh claimed entitlement only to her sexual freedom and her right to marry—not to her right to reproduce.

The HBF, Dickinson, and the California Department of Institutions agreed. "Had it not been for sterilization," Butler of Sonoma remarked, "It would not have been possible for these individuals to live as you and I desire to live, and as these patients are living today," referring to those he had permitted to marry.[115] G. B. Arnold, a physician at the Virginia State Colony for Epileptics and Feeble-Minded, concurred. "It is well known that many high-grade feebleminded women who are childless . . . have married and lived happily, and have successfully created homes for themselves and their husbands" after being sterilized.[116] Marriage, eugenicists believed, would provide a check on female promiscuity. As long as they could channel female sexual pleasure into marriage (and, for select women, into motherhood), civilization would advance. Neither moral decay nor racial degeneracy would plague the institutions of marriage and family. Sterilization, which restricted motherhood, did not take away sexual pleasure and did not increase premarital promiscuity (for, among other reasons, sterilized women could be released from institutions and allowed to marry). It offered a solution palatable to a wide range of people because it recognized modern notions of sexuality without necessarily promoting

[handwritten: fertility control by AMA, not personal choice]

reproductive rights (to which organizations such as the AMA were initially opposed). It approached the issue of fertility control in a scientific manner by emphasizing racial health, not individual choice.

More than just the ideas and policies promoted by a few individual fanatics, eugenic sterilization became in the 1920s a powerful movement with many public supporters. Those who expressed their concern to the HBF referred not only to themselves and their prospective mates. Many were eager to point fingers at family members, neighbors, or classmates. Like the parents who consented to—or even requested—having their daughters sterilized at the Sonoma State Home, others wrote the Foundation with specific individuals in mind. For example, an elderly man wrote expressing concern about his "happy but active granddaughter, about 16 years old whom I have long thought should be sterilized. She is by no means an imbecile on the contrary quite active and useful physically, goes to school and is a favorite everywhere. But she ought not to bear children," he declared.[117] Another wrote: *[handwritten: whose decision?]*

In my class in marriage relations I was given one of your enlightening publications. Just lately I learned that a baby was born to a girl who had been in grammar school with me. Frankly, that girl should not have had that child and should not have any more. She is feeble-minded and I understand her husband is also. In grammar school she could never learn. Finally, because of her large size and her age she was allowed to go to high school. I don't think she ever completed four years. Is there any way in which that couple could be legally sterilized? They are too ignorant to successfully use birth control methods. After going thru the Stockton State Hospital this week with a psychology course I am very anxious to prevent children from being born whose ultimate goal is such an institution. May I have some pamphlets to distribute in my home town?[118]

With the prodding of Dickinson, Laughlin, and the HBF, sterilization gained credence as a legitimate and ethical procedure for preventing procreation. Portraying the procedure as a harmless and effective remedy to improve the race, sterilization advocates gained a wide array of supporters—including patients, their family members, and their neighbors—who perceived the policy as not only effective but humanitarian. An Illinois history professor wrote Gosney expressing his sympathy for "liberal tendencies and programs for bettering the human race."[119] In another case, a woman discovered HBF publications in a Chicago medical library; she was inspired by their message and hoped to found a Chicago branch of the California-based Foundation. "Sterilization of the unfit goes hand in hand with Eugenics, and while we of this century can only lay the foun dations for it, those of humanitarian impulses, in years to come will build the structure, so that ultimately mankind will arrive at its ideal—a perfect mind in a perfect body—made possible by the propagation of the race by those who are mentally and physically fit."[120]

Thus, concern over racial health was no longer limited to "feeblemindedness." Beginning in the 1930s, eugenicists emphasized physical as well as mental health. This vision of a "perfect mind in a perfect body," expressed with increasing frequency in the 1930s and 1940s, pointed a finger at those whose "flaws" (large ears, abnormal height or weight, poverty, sexual promiscuity, or race) threatened to stand in the way of progress. Even religious figures sometimes perceived eugenics and sterilization as a moral benefit; for example, a rabbi at Temple Sinai in Oakland declared that Gosney's work was of "tremendous benefit to mankind. I am greatly thrilled at the thought that our own state, California, is pioneering in sterilization work," he continued. "Whenever I visit our state institutions—and I spent two days at the [Sonoma] Home for the Feebleminded at Eldridge last month—I always encourage the superintendents to continue their magnificent efforts toward improving the generation of tomorrow."[121]

Reflecting back on the sterilization movement, Dickinson noted its contributions to tomorrow's generation. Though active in other movements, such as infant and maternal health and birth control, Dickinson claimed of sterilization that "more striking results have been brought forward than for any movement of equal importance during my six decades of medical perplexities."[122] During the 1920s, sterilization advocates successfully appealed to the medical establishment and to the masses to support eugenic sterilization—an achievement overlooked by most historians. The 1930s would yield additional successes for the eugenic-sterilization movement as it became even more directly linked to "improving the generation of tomorrow." The notion of reproductive morality, developed by eugenicists to generate increased social concern about reproduction and racial health, took even greater hold on Depression-era America, when sacrificing individual liberties for the common good became a common and accepted concept.

CHAPTER FOUR

A New Deal for the Child

Ann Cooper Hewitt
and Sterilization in the 1930s

[handwritten: concurrent with depression — priviledged problem to be priny to]

From January through August of 1936, the *New York Times*
offered readers nearly fifty articles on a sensational court case taking place
twenty-five hundred miles away in the city of San Francisco. In this case,
which received coverage around the globe, twenty-one-year-old Ann
Cooper Hewitt, heiress to millions from her late father, Peter Cooper
Hewitt, sued her mother and two surgeons for surgically sterilizing her
without her knowledge. At stake was not only the legality of sterilization
but the meaning of eugenics and the appropriate balance of power between mothers and daughters in 1930s America. The Cooper Hewitt trial
signaled a new phase in eugenic strategies, which focused on the importance of motherhood and family to the future of the race. In addition,
the widespread coverage of the trial moved eugenics more directly into
the public eye, generating greater discussion and acceptance of sterilization as a social stabilizer in a turbulent decade. During the 1930s, eugenics played a formative role in the transition from a culture of individualism to a culture of responsibility by offering sterilization as a way to
strengthen the American family.[1] The Cooper Hewitt story helps to explain how eugenics actually gained influence in a decade in which the
movement was previously believed to have been dismantled.

In the 1930s, the nuclear family became a central subject of public policy and popular debate. Scrutinized by sociologists and eulogized by New
Dealers, the American family emerged as an institution essential for surviving the Depression but also threatened by it. Many Americans noted
that the economic crisis had a destructive impact on the family.[2] When so-

[handwritten: did people not having children in Depression era have to do with money or eugenics?]

ciologists Robert and Helen Lynd returned to their study of "Middle-town" (Muncie, Indiana) during the Depression, they noted increased tensions in marital relationships that had not been evident in the 1920s. Husbands and wives sometimes reacted to economic pressures with mutual recrimination. "Have you anybody you can send around to tell my wife you have no job to give me?" one husband asked a social worker. "Certainly I lost my love for him," reported a wife to another social worker. "How can you love a husband who causes you so much suffering?"[3]

Yet sociologists also reported that the Depression had brought the family closer together. One noted that "many a family that has lost its car has found its soul." The Lynds were heartened by an editorial in a Middletown newspaper that suggested the ways in which the economic situation worked to strengthen family values. "All of us are hoping for a quick return of the prosperity we once knew," the editorial remarked, ". . . but in the meantime, some millions of Americans already have a kind of prosperity that includes the strengthening of family ties, better health, and the luxury of simple pleasures and quiet surroundings."[4] With so much invested in the family as a "spiritual guide to survival in hard times," its preservation and stability became crucial to many Americans.[5]

Indeed, the Depression severely strained gender and family relations. By 1933, 25 percent of America's workers were unemployed. Average family income decreased 40 percent, from $2,300 in 1929 to $1,500 four years later. Unemployment and lowered living standards undermined men's self-esteem and resulted in a high rate of desertion. By 1940, over 1.5 million married women were living apart from their husbands.[6]

Marriage rates declined as well; in 1932 there were 250,000 fewer marriages than in 1929, and a total of 800,000 marriages were postponed because of the Depression. As Caroline Bird recalled, most young couples worried about whether they could afford to get married on $15 a week, and many did not even have that much. One Chicago schoolteacher noted the impact of the financial constraints on young couples during the 1930s. "Do you realize how many people in my generation are not married?" she remarked. "It wasn't that we didn't have a chance. I was going with someone when the depression hit. We probably would have gotten married. He was a commercial artist and doing very well. . . . Suddenly he was laid off. It hit him like a ton of bricks. And he just disappeared."[7] As a result of the Depression, more women found themselves single or living independently from their husbands than ever before.

Though the number of marriages dropped precipitously during the 1930s, the most dramatic decline was in the birthrate, which for the first

↓ marriage.   ↓ birthrate

time in American history dropped below the replacement level. During the 1930s, Americans produced nearly three million fewer babies than they would have had at the 1929 rate. Most middle-class couples limited their families to two children. A familiar cliché of the 1930s was "I don't want to bring children into a world that has no use for them."[8]

Adding to the accelerating problem of a declining birthrate was the growing recognition that poorer women were producing more children than wealthier women. Feminists and liberals pointed out that half of all births during the 1930s came from families on relief or making under $1,000 a year. This information reinforced the eugenic concern over racial degeneration. Civilization would not advance if the weaker, more dependent members of society produced the greatest number of children. As a columnist for the *Los Angeles Times* remarked in 1936, "In poverty the breeding is prolific; and it is indisputable that the great majority of the feebleminded are impoverished and on relief." A statistician for the Metropolitan Life Insurance Company also observed a connection between poverty and procreation, remarking that "it is always the least desirable parents who are the last to curtail their fecundity."[9]

Thus eugenicists shared the widespread concern over the destructive impact of the Depression on the American family. Yet they believed that the Depression did not cause the decline of the family but merely revealed how unstable it had become. They argued that promiscuity, not poverty, was at the root of family pathology. The culture of the 1920s, which validated female sexual pleasure and female independence, had led to a widespread desertion of domesticity. As one eugenicist explained in a 1930 address at the University of California, Berkeley, women were destroying civilization by "turning away from their natural modes of expression in the home and family."[10] With illegitimacy, premarital sex, and divorce all on the rise, eugenicists believed the traditional family model was destined for extinction. Sterilization, they argued, could prevent such extinction. A policy of selective sterilization would guarantee that only those demonstrating the morality and ability necessary to raise a child would be allowed to do so. Every child would be born with a sound mind in a sound body, and both family and civilization would thus be stabilized and strengthened.

The family also gained the attention of academia. The rise of the social sciences in the late 1920s and 1930s and the formation of the Social Science Research Council in 1923 legitimized the work of sociologists and anthropologists concerning the nature and origins of human behavior.[11] Emphasizing culture (or nurture or environment) over biology (or nature or heredity), these scientists succeeded by 1930 in displacing the belief in the biological basis of human behavior that had dominated academic institutions in the 1920s. By shifting the "causal link between parents and children" from heredity to environment, the social sciences located the origin of social pathology in an individual's culture rather than in his or her genes. In this context, the preservation of the nuclear family and the parent/child relationship gained even greater significance; personality was shaped more by family relationships than by genetic makeup.[12]

Redefining Eugenics

This newfound emphasis on environment over heredity in the social sciences has misled many historians to dismiss the role of eugenics in the 1930s. Assuming that eugenicists were concerned exclusively with heredity and disregarded the role of environment, scholars have concluded that eugenics lost all scientific and academic credibility and support in the 1930s. An influential authority in the history of eugenics declared in 1963 that "the most significant development in eugenics after 1930 was its rapid decline in popularity and prestige," a claim that has remained largely unchallenged.[13]

As a result, U.S. historians have ignored the role of eugenics in a significant decade, one in which eugenicists promoted sterilization as a panacea for the problems of the family. Specifically, eugenicists popularized a doctrine of reproductive morality that countered selfish individualism with social responsibility. They transformed the politics of reproduction from a private matter of personal liberty to a public issue of racial health and, with the assistance of the widely publicized Cooper Hewitt trial, convinced the public that sacrificing reproductive freedom for the sake of stabilizing the American family was well worth it. In this sense, eugenic sterilization became the social security of American civilization: it guaranteed a healthy and stable future citizenry. It promoted the idea of increasing state intervention in and regulation of previously private matters for the protection and stabilization of social institutions—the very model on which Franklin Roosevelt would construct his New Deal policies. The familial language which eugenicists appropriated in the 1930s proved potent. The combined ideal of strengthening the race and preserving the family held great public appeal and paved the way for public acceptance of New Deal social liberalism and its "sacrifice of liberty for the sake of stability." Americans were not "dragged kicking and screaming toward social responsibility," as historian Alan Dawley argues; they had already accepted it as an essential step in protecting and preserving the American family.[14]

Eugenicists' emphasis on the family and on the importance of motherhood was also a survival strategy. As a result of the social sciences' emphasis on environment, as well as genetic research, eugenics and sterilization did come under attack in the 1930s. For example, Johns Hopkins biologist Herbert Spencer Jennings argued that even if all feebleminded Americans were prevented from reproducing, it would take sixty-eight generations, or two to three thousand years, to decrease the proportion of feebleminded in the population to one per ten thousand. The problem, he argued, was that "normal" people also could be carriers of feeblemindedness. A professor of pathological history, Alexander Fraser, elaborated on Jennings's argument, declaring in 1934 that sterilization of the feebleminded was "a biologically unsound procedure" because feeblemindedness was often a recessive gene: healthy people could carry the defect and never know it until they had passed it on to a child. Citing geneticist Thomas Hunt Morgan's work on the fruit fly, which revealed the complexity of chromosomes, Fraser pointed out that it would be impossible to detect potential carriers of feeblemindedness, as such people were "normal, or even superior, in their phenotypical characteristics." Furthermore, Fraser declared, eugenicists ignored the importance of the environment in the development of individuals. For these reasons, he recommended that sterilization be stopped.[15]

Thus, eugenics was gaining enemies from both sides: biologists and geneticists, who found the movement's science overly simplistic, and social scientists, who deplored its ignorance of culture. Historian Carl Degler concludes that "the death knell for eugenics" was sounded not by anthropologists and sociologists but by geneticists "as they learned that their science was being both misunderstood and misused by public advocates of eugenics." Once biologists and geneticists dismissed eugenics, he argues, the social sciences followed suit. As a result, "by the 1930s the popularity of eugenics declined precipitously, not only among social scientists and biologists but among the public."[16]

But historians such as Degler and Mark Haller misread the impact of the attack on eugenics. Accepting the claims of the opposition at face value, they assume such denunciations of eugenics convinced the public of its fallacy—while providing no evidence that it did so. They overlook the fact that the response and recovery of eugenicists did not in fact di mantle their program but bolstered it. Though they are forced to acknowledge that the practice of sterilization increased significantly during the 1930s, historians either regard this increase as a cultural "lag" or portray sterilization as an "extremist cause separate from more mainstream eugenic thinking" that "assumed a life of its own."[17]

In the face of attack, sterilization advocates creatively adapted their definition of eugenics. No longer pitting nature against nurture, they combined both in their new formula, which was centered on the importance of motherhood and the family. In order to maintain scientific legitimacy, they emphasized the importance of the home environment for child development. This new strategy, which shifted the emphasis from heredity to maternal care, worked to strengthen and consolidate their movement. For example, though Sonoma State Home superintendent Fred Butler had previously considered "the heredity type of individuals" first when deciding which cases to accept, he noted that by 1930 "we [had] reached the point where we *practically disregard*[*ed*] whether they are the hereditary or non-hereditary type." In either case, the patient would not make a suitable mother because she would not care for the child properly. "The saddest of them all, in my opinion," he wrote in 1931, "is a feebleminded mother trying to care for normal children."[18] Regardless of what ailed his patients, they would not make suitable mothers.

By emphasizing environment rather than heredity, the eugenics movement survived the attack by geneticists and social scientists and flourished in a society in search of immediate and effective solutions to severe economic and social problems as well as for ways to stabilize the family. Touting sterilization as a cost-effective strategy for the production of "better" children and the rehabilitation of the American family, eugenicists attained a greater degree of cultural authority than ever before. By focusing on sterilization as a means of restricting motherhood rather than of eliminating genetic defects, eugenicists escaped the limitations of hereditary arguments. The creation of a better race would come from the production of better children, they argued in the 1930s—a catchy idea that inspired many, including President Herbert Hoover, to pay increased attention to eugenics. "There shall be no child in America," Hoover declared, "that has not the complete birthright of a sound mind in a sound body." (This statement, known as the Child's Bill of Rights, inspired the name of a national sterilization organization, Birthright.)[19] If environment was now as important a consideration as heredity, then it was all the more important for society to exercise reproductive morality—to be selective about who should raise America's children. Motherhood was not simply a personal choice; it was a political act that many believed should require proof of merit.

This new focus on tomorrow's children rather than on today's "misfits" was an attempt to modernize the movement in order to enhance its appeal. Because children and family gained media coverage as well as public and professional interest during the Depression years, eugenicists incorporated these popular family issues into their public appeal to stabilize the race. Some active sterilization advocates, such as Paul Popenoe and Lewis Terman, embarked on marriage and family studies in the 1930s in an attempt to widen their eugenic influence. Others remained focused on sterilization but emphasized its benefits to children and family. Regardless of which policies they advocated, eugenicists in the 1930s were careful to acknowledge the role of environment in the development of a child's behavior.

Eugenicists had the opportunity to influence popular attitudes and public policy toward children beginning in 1930, when they played a leading role in President Hoover's White House Conference on Child Health and Protection, the first Executive-inspired event on child development since 1909.[20] Four thousand professionals attended the Washington conference, which presented the investigative work of twelve hundred specialists, including physicians, psychologists, home economists, sociologists, and eugenicists. One hundred and seventy-five committees issued a total of 141 reports on "every angle" of the well-being of the child, from public health, education, and child labor to treatment of the mentally deficient and delinquent. Significantly, the conference chair, Secretary of the Interior Ray Lyman Wilbur, was an active member of the American Eugenics Society. He declared the conference a "far-seeing move in the interest of our future citizenry." Applying eugenic language to the issue of child welfare, Wilbur explained that the conference's ultimate goal was "making this a fitter country in which to bring up children."[21]

President Hoover's vision of the future also incorporated a eugenic message. On November 19, in a live radio address, Hoover confronted the nation about the need for healthy children to stand up under the "increasing pressure of life." Hoover used the image of the physically and mentally strong child as a symbol of America's potential. Addressing the "unseen millions listening in their homes" (more than twelve million families owned radio sets in 1929),[22] he explained that the purpose of the White House conference was to provide parents with information and safeguards to ensure their children's "health in mind and body," a measure designed to guarantee future success. "If we could have but one gen-

eration of properly born, trained, educated and happy children," he explained to families listening at home, "a thousand other problems of government would vanish. We would assure ourselves of healthier minds in
more vigorous bodies, to direct the energies of our nation to yet greater
heights of achievement."[23] Implicit in this statement was the necessity to
eliminate the unhealthy, whose burdens stood in the way of cultural
progress.

In stark contrast to this image of physical and mental perfection,
Hoover cited statistics intended to shock his listeners. Of the nation's
forty-five million children, ten million suffered some sort of mental or
physical deficiency and therefore hindered human progress. It was not
the poverty of the Depression that prevented America's children from
embodying physical and mental ideals, Hoover argued, but "ignorant
parents" and "ill-instructed children." As the *Survey* argued the following
year, the White House Conference had "put the parents of the United
States on trial." Though the government had made progress in eliminating deficiency (or at least in making it invisible), having "cleared our
streets and schools of the hump-backed, the crooked legged," it was up
to parents, particularly mothers, to ensure that their children were born
and raised properly.[24]

Yet experts, including eugenicists, were available to guide them. A
"growing battery of professionals" such as those present at the Conference joined mothers in supervising social reproduction "from cradle to
grave."[25] Among them, eugenicists such as Popenoe and Sonoma State
Home Medical Superintendent Butler played a role as members of the
White House Conference Committee on Mental Deficiency (Butler was
the chairman of the California section on mental deficiency). The Committee, defining mental deficiency as feeblemindedness and intellectual
subnormality, found that 13 percent of the total population suffered from
subnormal intellect and an additional 2 percent were "definitely feebleminded" and "socially inadequate."[26]

The solution to this crisis, the Committee concluded, was twofold:
foster increased social responsibility for the mentally deficient and promote sterilization of the unfit. Sterilization, the Committee suggested, if
used selectively (that is, on a case-by-case basis rather than on all members of a racial, ethnic, or gender group), would effectively curtail the
problem of mental deficiency. Undoubtedly influenced by the research
of Popenoe and the practice of Butler and Dickinson, the Committee supported the use of eugenic sterilization, declaring, "There is now little
doubt that properly drafted laws permitting sterilization are constitu-

tional, and that the operations necessary (vasectomy and salpingectomy) cause no recognizable alterations in personality and are not attended by appreciable mortality."[27] The Committee convinced the White House Conference to go "squarely on record in favor of sterilization."[28]

Hoover's White House Conference and the conclusions of its Committee on Mental Deficiency thus promoted sterilization as a social stabilizer to the American public. As Hoover suggested, civilization and the family would only march forward "on the feet of healthy children."[29] If reproduction were restricted in the name of social responsibility, healthy children would ensue.

Three years after Hoover's conference drew national attention to the problem of "defective" children, sterilization achieved international repute. In 1933, shortly after Hitler came to power, the Nazis enacted a eugenic-sterilization law that targeted the mentally and physically deficient and ultimately resulted in the sterilization of more than 350,000.[30] Initially, American eugenicists were delighted with Hitler's policies. Indeed, California eugenicists were actively involved in the enactment of Germany's sterilization law. Gosney and Popenoe corresponded regularly with many German eugenicists and sent them their publication on California sterilization, *Sterilization for Human Betterment*. Eugen Fischer, the director of the Kaiser Wilhelm Institute for Anthropology, Human Heredity, and Eugenics in Berlin, thanked Gosney for the book, which he found fascinating. "The proof that the sterilized people did not suffer any individual damage seems of particular importance to me," he wrote in February 1930. "Particularly these issues will be of great importance for legislation here in Europe."[31] Dr. Hermann Simon, director of another eugenics institute in Gütersloh, also expressed enthusiastic interest in *Sterilization for Human Betterment* and remarked in 1930, "I do hope the time will come, that also in Germany this knowledge will be generalized."[32]

Just three years later, Simon's dream became a reality. Germany enacted its sterilization statute, based on that of California, and many eugenicists in America initially followed that country's policies with enthusiasm. Dickinson, whose own interest in sterilization had led him to study the procedure in California institutions in the 1920s, sent a researcher to Germany in 1935 to study its policy. She reported back that "the leaders in the sterilization movement in Germany tell one over and over again that their sterilization legislation was formulated after careful study of the California experiment under Mr. Gosney and Dr. Popenoe's leadership. It would have been impossible, they say, to undertake such a venture involving two million people without drawing heavily upon previous experience elsewhere."[33]

Popenoe and Gosney defended the German law through the 1930s. In both private correspondence and publications, Popenoe emphasized that the law was "not [a] hasty improvisation of the Nazi regime" but the product of years of sterilization practice and research. Indeed, Popenoe believed that Hitler's rise to power merely ensured that these ideas would be carried out.[34] In an article published in the *Journal of Heredity,* Popenoe praised the German government for developing a solid eugenic policy that appeared to "accord with the best thoughts of eugenists in all civilized countries." In Popenoe's estimation, Hitler's ideas about human progress and the advancement of civilization were no different from those popularly expressed in the United States. Drawing from Hitler's *Mein Kampf* (1923), he quoted a passage remarkably similar to Hoover's Child's Bill of Rights. "He who is not sound and worthy in body and mind should not perpetuate his handicaps in the bodies of his children," declared Hitler in the volume.[35]

By 1936, California eugenicists recognized that Hitler's persecution of the Jews might undermine the credibility and support of eugenics. Some continued to believe, however, that Hitler would be remembered not for his "political high crimes" (which would soon be forgotten) "but as the first head of a modern government who enforced legislation for the elimination of the unfit for the biological improvement of the race." *Los Angeles Times* columnist Fred Hogue quoted a "venerable student of history of international repute" as saying that "the evolutionary development of the race is much more important than passing political and social revolutions."[36]

But, by the end of the decade, Gosney and Popenoe wanted to avoid any associations with Hitler, who had little popularity in the United States. They avoided the racial categories used by Hitler and suggested that sterilization in America would be used selectively rather than on large groups. In 1940, Gosney warned his staff to avoid using racial terms in HBF literature. "We have little in this country to consider in *racial integrity*. Germany is pushing that. We should steer clear of it lest we be misunderstood."[37] Popenoe wrote to a colleague in 1945, "When it comes to eugenics, the subject of 'race' sets off such tantrums in a lot of persons that one has to be very long-suffering!" Popenoe concluded that it was best to avoid the term altogether (though he and all eugenicists continued to use the term "race betterment" when referring to the goal of eugenics).[38]

Hitler was not the only inspiration for the new terminology of "selective sterilization" in the American eugenics movement. The growth of mass culture was blurring racial distinctions by the 1930s. Mass consumption and the development of national chain stores, brand names, and mail-order catalogs, as well as new forms of communication, forged a new era of cultural homogeneity that glossed over the previous era's emphasis on racial differences.[39] Though America was more culturally diverse in the 1930s than in any previous decade as a result of massive immigration between the 1880s and 1920s, modern mass culture exposed the population "to unprecedented pressures toward cultural uniformity."[40] In this modern culture, normality—that is, the concept of a culturally constructed, gender-specific standard of behavior, ability, and appearance—became increasingly prevalent as a key principle in American society.[41] Abnormality, rather than race, class, or ethnicity, differentiated those who, from a eugenic perspective, should not reproduce from those who should.

As a result, eugenicists used the term *selective sterilization* to explain the goal of eugenics: civilization would improve when only those considered "normal" were allowed to reproduce. Sterilization promised "sound minds in sound bodies" to fit the images promoted in Hoover's political speeches as well as the increasingly rigid and standardized images of beauty promoted by mass culture.[42] Dismissing explicit racial categories, selective sterilization targeted abnormality—a flexible category of physical and mental traits that allowed for both hereditary and environmental causes of difference.

For example, in 1937, Frederick Osborn, president of the American Eugenics Society, stressed the importance of avoiding wholesale categorization of sterilization candidates. Desiring to distance American eugenics from German eugenics, he warned that it "would be unwise for eugenists to impute superiorities or inferiorities of a biological nature to social classes, to regional groups, or to races as a whole. . . . Eugenics should therefore operate on the basis of individual selection." He continued, "Fortunately, the selection desirable from the point of view of heredity appears to coincide with the selection desirable from the environmental point of view."[43] Thus, the term *selective sterilization* enabled eugenicists to target individuals based not on the cause of their supposed deficiency but on their desirability as a parent. Whether they emphasized selective or eugenic sterilization, heredity or environment, their goal remained consistent: to develop a mainstream following that believed in the concept of reproductive morality and accepted its mandate to sacrifice personal liberty in order to strengthen family and community.

Thus, events in the late 1920s and early 1930s—the rise of the social sciences and the resulting emphasis on environment, discoveries in genetics, and the Nazi use of sterilization—forced American eugenicists to justify sterilization on new grounds. Most important, eugenicists stressed that sterilization should be used to weed out those, usually women, who would not make capable parents. As the president of the American Association of Mental Deficiency argued in 1936, "Probably the most powerful argument for sterilization today is that which urges that no feeble-minded person is fit to be a parent, whether or not his condition is hereditary and therefore likely to be genetically transmitted."[44]

Beginning in the 1930s, then, eugenicists who had previously established their careers on the principles of heredity seemingly contradicted themselves by inviting environmental factors to come into play, a move that previous historians have interpreted as an indication of eugenic defeat. But, ultimately, this was a smart tactic. It saved the movement from extinction, and it also widened eugenicists' sphere of influence and further popularized their goal to improve civilization by making reproduction a social and medical responsibility rather than an individual right.

As a result, sterilization gained supporters in the 1930s. Incorporating an environmental justification for sterilization paved the way for its increased use. Dr. Gladys Schwesinger, addressing the New Jersey Health and Sanitary Association in 1937, emphasized the "right of every child to have competent parents." Playing on Roosevelt's New Deal policies, she argued that sterilization played a key role in "procuring a better deal for the child." Children should not be entrusted to just anyone; the same supervisory control should be administered over parenthood as over teaching, nursing, and other services. Given that a child's personality could be "made or marred" at home, it was all the more important to protect children from "the wrong kind of parents." Sterilization would spare "many unborn children the affliction of being born to unqualified parents." Echoing the eugenic emphasis on reproductive morality as an essential component of modern society, Schwesinger pleaded for Americans to heed the call of social responsibility. "It is society's obligation," she argued, "to encourage the burden and responsibility of parenthood in the best and withhold it from the worst individuals of each generation."[45]

Schwesinger's call for social responsibility reinforced a central aspect of Roosevelt's New Deal policy, established when he assumed the presidency in 1933. For the first time in American history, the federal government assumed responsibility for family welfare. Whereas Hoover had intimated that mothers needed the aid of experts in raising physically and mentally healthy children, Roosevelt suggested that the federal government itself should become more involved than it had been in the lives and welfare of the American family.[46]

Even prior to New Deal politics, eugenicists announced that it was time for the government to aid their cause. "Race Degeneration Seen for America," the *New York Times* announced in June 1932 in reference to an HBF paper presented at the American Association for the Advancement of Science. If the government did not intervene, eighteen million people burdened by disease or mental defect would become a charge and burden on the rest of the population, the Foundation declared. Emphasizing the importance of collective health over individual rights, the Foundation demanded that the government take an increasingly active role in regulating fertility. "The law of self-preservation is as necessary for a nation as an individual," the paper announced.[47] Certain women should not have children, and the government should expand and enforce sterilization laws.

Legislators heeded the eugenicists' call. Between 1931 and 1939, over twenty thousand institutionalized patients were sterilized, nearly triple the number sterilized between 1920 and 1929.[48] Seventeen states enacted or revised sterilization laws during a four-year period after the Supreme Court upheld the constitutionality of Virginia's involuntary sterilization law in the case of *Buck v. Bell* in 1927.[49] By 1936, Delaware, Idaho, Iowa, Michigan, North Carolina, Oregon, South Dakota, and Vermont had all expanded their statutes to allow the sterilization of those mentally deficient who were not institutionalized. Nebraska instituted a law in 1935 requiring the registration of all feebleminded in the state and denying a marriage license to any such person without proof of sterilization. As legal analysts underscored, heredity was no longer a necessary element of eugenic-sterilization laws. Nebraska's law was "based mainly on a consideration of feeble-mindedness from the social point of view," Popenoe observed. "No mention is made of any considerations regarding the transmission of defective genes to offspring." Parenthood was becoming an increasingly exclusive privilege; as Popenoe noted of the Nebraska law, "Though some feebleminded parents might produce a small proportion of normal children, the Nebraska legislators evidently did not think it desirable to have even normal children brought up by feebleminded parents."[50] Thus, laws originally intended to prevent the hereditary transmission of unfavorable genetic traits were bolstered and more widely utilized after belief in their original purpose had virtually disappeared.

In this diffuse climate of anxiety about the future of tomorrow's children, a court case erupted that would change the way many Americans thought about sterilization and family politics. As Morantz-Sanchez

Doctor Upholds Mental Fitness Of Miss Hewitt

1 — 9 — 1936

Jersey State Hospital Official Says November 7 Tests Showed Her Alert

Criticizes Environment

Mother Admits Operation, Defends It as Kind Action

Special to the Herald Tribune

HACKENSACK, N. J., Jan. 8.—Miss Ann Cooper Hewitt, said by surgeons in San Francisco to have been subjected to a sterilization operation because she was feeble-minded, submitted to a mental and physical examination last November 7, by Dr. Lawrence A. Collins, senior resident physician of the New Jersey State Hospital for the Insane, at Greystone Park. He found that she "does not present any intellectual defects or mental aberrations and is not suffering from any mental diseases."

While this was revealed in an affidavit filed in Chancery Court here, a statement was issued through a lawyer by Mrs. Maryon Hewitt McCarter, the girl's mother, commenting on Miss Hewitt's $500,000 damage suit against her in San Francisco and the charge that she subjected the girl to sterilization.

Mother Defends Action

Admitting that an operation had been performed, the statement said

She Accuses Her Mother

Herald Tribune photo—Acme
Ann Cooper Hewitt

Criticizes Early Environment

"In view of the history and of the result of my examination, it is my belief that this young girl has been conditioned during her early formative years by an unwholesome environment, and that any intellectual deficiency which might be present is due not to any pathological defects but to the lack of development of her intellectual faculties," said the Collins affidavit.

Mr. Breslin, in his statement for the girl's mother, referred to her as Mrs. Hewitt. She was divorced last

Figure 5. Ann Cooper Hewitt, *New York Herald Tribune*, January 9, 1936.

Figure 6. Mrs. Maryon Cooper Hewitt, *New York Herald Tribune,* December 8, 1936.

argues, sensational trials simultaneously tell two stories: one pertains to the trial participants; the second, "perhaps even more compelling and instructive than the first, mirrors the wider field of politics and culture in which the trial takes place."[51] Both these stories were told in the Cooper Hewitt case. When the wealthy Ann Cooper Hewitt (figure 5) took the stand in San Francisco in 1936 to testify that she had been sterilized without her knowledge or consent, Americans followed her story with fascination. She was not an institutional inmate, and she was far from poor. Her case symbolized the new role that sterilization played in the 1930s as a way to strengthen the American family.

Newspaper coverage of the case emphasized Ann's relationship with her mother, who had ordered the operation. The behavior of Mrs. Maryon Cooper Hewitt (figure 6) was also on trial, as journalists and observers attempted to determine whether she was an irresponsible mother.

The *Literary Digest* labeled the case the "daughter-versus-mother drama."[52] Unlike any sterilization event before it, the Cooper Hewitt trial riveted the American public. Even the well-known *Buck v. Bell* Supreme Court case (upholding the constitutionality of eugenic sterilization), which historians point to as a highly significant event in the history of eugenics, failed to capture public attention and stimulated no editorial comment;[53] the *New York Times* did not print a single article on the case. Ann's story succeeded in capturing public attention because it addressed the very issues eugenicists had put forward in their campaign of the 1930s: the relevance and legitimacy of environmental factors in determining who should be sterilized and the importance of protecting the child and preserving the family.

In addition, the trial drew public attention because of the legal questions it raised about both sterilization and sexuality. Because Cooper Hewitt was sterilized by a private physician rather than an institution, her case hinged on the legality of sterilization in private medical practice. As the California sterilization law did not specify whether such practice was allowable and no previous court decisions had established a precedent, eugenicists hoped that the Cooper Hewitt case would clarify the law and increase the number of sterilizations performed in private practice. Female sexuality was also on trial in the case; legal analysts debated whether the primary function of sexuality for women was procreation or pleasure. The surgeons were charged with "conspiracy to commit mayhem" on the grounds that sterilization fell into this category. The crime of mayhem consisted of "unlawfully and maliciously depriving a human being of a member of his body, disabling, disfiguring, or rendering it useless."[54] Yet, as Dickinson, the HBF, and other eugenicists had effectively argued, sterilization did not "unsex" the patient, as she could still experience sexual pleasure. Also central to the case, then, was the lingering question of the meaning of female sexuality in American society.

The Cooper Hewitt story, debated by legal experts, sensationalized by journalists, and followed by millions, introduced sterilization as a family-centered solution to the problem of female sexuality. Newly packaged with enhanced appeal to a 1930s audience, eugenics addressed issues relevant to a majority of Americans: the survival and stability of the family as a social institution, the usefulness of "normality" as a central organizing concept, and the importance of environment (in addition to heredity) to human development. Portrayed in a eugenic narrative, fertility left unchecked had dangerous implications for both individual families and American civilization. By linking the personal to the larger social body in their campaign, eugenicists raised the stakes of procreation and drew Americans toward their doctrine of reproductive morality and its message of social responsibility.

Sterilization on Trial

On January 6, 1936, in the city of San Francisco, Ann Cooper Hewitt filed a $500,000 damage suit against her mother and two surgeons who had sterilized her without her knowledge in August 1934. She had entered the hospital just eleven months short of her twenty-first birthday for an emergency appendectomy, but, after the administration of a mental test that labeled her a moron and at the request of her mother, she received a salpingectomy as well. Charging that the operation was part of a plot to deprive her of her inheritance, she sought remuneration from those responsible. In the will of her late father, Peter Cooper Hewitt, Ann was to receive two-thirds of the millionaire's estate, but if she died childless, the estate would revert to her mother. She believed, therefore, that her mother had requested the operation in order to ensure that Ann would lose her fortune. Though the case initially appeared to be concerned with financial matters, it was quickly shaped into a story about sterilization and the family.

Ann's charges led to a criminal investigation the following day, and the case quickly took on increased significance as it raised questions about the California sterilization law and about parental power over minors. Based on Ann's testimony, the San Francisco municipal court charged her mother and the two surgeons with "conspiracy to commit mayhem" and issued warrants of arrest based on invasion of personal rights.[55]

After filing an affidavit denying any wrongdoing, Ann's mother fled to New Jersey, attempted suicide, and spent five months in the psychopathic ward of the Jersey City Medical Center, distraught over her daughter's accusations. Though she was indicted, the San Francisco district attorney informally dismissed the charges in December 1936 because of Ann's reluctance to testify against her mother.

California surgeons Tilton E. Tillman and Samuel G. Boyd did not escape a trial. After months of discussion and preparation, the case went before Superior Court Judge Raglan Tuttle in August 1936. For six days, a jury listened to family members, expert witnesses, and the arguments of prosecution and defense. Finally, Tuttle dismissed the case and all charges against the physicians on the grounds that sterilization was perfectly legal and that mayhem had not been committed. Though the prosecution filed an appeal, the State Supreme Court declined to reopen the case. To the delight of sterilization advocates, the Cooper Hewitt case ended in their favor by validating the use of sterilization in private practice even without the consent of the patient.

Though Ann's mother was noticeably absent from the August proceedings involving Tillman and Boyd, coverage of the case still centered on her and her relationship with Ann. Between January and August of 1936, family members, health professionals, hospital attendants, and other witnesses, along with the press, constructed two competing narratives of the mother-daughter relationship. In one, Mrs. Hewitt was a conscientious, indulgent, and loving mother who sought only to protect society and a "dangerously oversexed" daughter from her own mental and moral incompetence. In the other, Mrs. Hewitt, five-time divorcee, was an abusive mother who deprived her neglected-but-otherwise-normal daughter of love, education, and her right to reproduce. The conflicting narratives engaged the American public because each offered a villain who possessed the symbolic power to destroy the American family. Either an oversexed daughter or an abusive mother needed to be reckoned with. By focusing on the construction and use of these competing narratives, we can explore the multiple meanings of the trial to those involved in and witnessing the drama.[56]

Both subjects were on trial for the same crime: their incapacity to mother. While the prosecution cited evidence that Mrs. Hewitt's inability to control Ann stemmed from her failure as a mother, the defense charged that Ann herself would never be a "desirable" or "competent" mother, thus justifying her sterilization. To fashion narratives they believed would most likely appeal to a jury of ordinary Americans, both sides placed their protagonists within the context of the family. As the case underscored, family pathology was not just a problem of the Depression-era poor. The real problem was not financial but feminine: both mother and daughter had deserted domesticity, violating their prescribed gender roles at a time when such prescriptions were deemed crucial to social survival.

When Ann's lawyer, Russell P. Tyler, announced in January 1936 that Ann was suing her mother, he called the mother's reputation into question by using her full name. The suit, he declared "will be against Miss Hewitt's mother, Mrs. Maryon Bruguiere-Denning-Hewitt-D'Erlanger-McCarter."[57] The press consistently printed her full string of names, thus emphasizing her five failed marriages. Though her marriages placed her in a position of great affluence and made her a socialite known in San Francisco, New York, London, and Paris, she had ultimately failed as a wife.[58]

In Tyler's portrayal of Mrs. Bruguiere-Denning-Hewitt-D'Erlanger-McCarter, she was no better at motherhood than she was at marriage. He charged that throughout Ann's childhood she had "failed in her duty as a mother and guardian" and "shamefully neglected" Ann's education. Though one of the wealthiest women in the United States, she deprived Ann of everything to which a financially privileged child was entitled. As a result, Ann was "virtually the prisoner of her mother," Tyler declared, "a victim of cruelty, forced to live with hardly more than the bare necessities or comforts of a poorhouse waif."[59]

In this narrative, Mrs. Hewitt's greatest transgression was the neglect of her child. During the Depression, sensitivity toward and public awareness of child neglect was on the increase. In an era that witnessed a return to traditional gender roles, "neglect became *by definition* a female form of mistreatment."[60] Neglected children signaled neglected domesticity—and, in the case of Ann's mother, evidence of one suggested the presence of the other. She was thus allegedly a failure as both wife and mother.

In a public statement made immediately after filing suit, Ann reinforced Tyler's portrait of her childhood. "I had no dolls when I was little," she began, "and I'll have no children when I'm old. That's my story." Described as "timid and nervous," she confessed to an unhappy childhood in which her mother frequently kept her "locked up" and called her an "imbecile."[61] Instead of being nurtured by her mother, she watched much of her fortune disappear, she claimed, as her mother had "squandered hundreds of thousands of dollars" of Ann's money at gambling resorts around the world.[62]

Three nurses who had attended her after the sterilization operation validated Ann's claims of abuse. Grace Wilkins recalled that when hired to take care of Ann, she was told that Ann was a "mental case." Yet after only a half hour in attendance, she believed that Ann suffered not from mental deficiency but from abuse. Over a period of three months, Wilkins witnessed Ann's letters and phone calls being censored by her controlling mother.[63] A second nurse, Anne Lindsay, assisted in Ann's sterilization but also found her to be "entirely normal."[64] Sarah Bradford, who cared for Ann during her month of convalescence, testified that her mother was "cruel, inhuman, unnatural and inconsiderate." Describing Mrs. Hewitt's "late night drinking parties," Bradford noted that she "did not conduct herself as a *normal* mother would."[65]

The sworn affidavits of Ann, her lawyer, and three attendants who had witnessed the mother-daughter relationship firsthand portrayed Ann as a timid, undeserving victim of her mother's greed and wrath. Emphasizing that Ann herself was "normal," they constructed a portrait of her mother as monstrous and inhuman, incapable of performing the duties of motherhood. Maryon Hewitt symbolized the poisonous effect of the 1920s culture of individualism on the American family. As an independent woman, traveler, drinker, and gambler who gave birth to her daughter out of wedlock (only afterward marrying Ann's father), she demonstrated to eugenicists and other social critics that the widespread desertion of domesticity in the 1920s had severe ramifications for the American family in the 1930s. In this narrative, it was she, not Ann, who posed the greater threat to the family's survival.

Though witnesses testified to Ann's normality, they lacked the scientific expertise to free Ann from accusations of abnormality. As eugenicists had argued since the turn of the century, the real danger of high-grade morons (as experts labeled Ann) was that they could pass as normal. Hewitt may have been able to fool those around her that she possessed normal mental faculties, skeptics argued. Evidence of her pathology stemmed from the report of psychologist Mary Scally, head of the Division of Mental Hygiene of the Board of Health of San Francisco. Hours before Ann's sterilization, Scally diagnosed her as a moron with a mental age of ten years and eleven months. "Had she applied herself more persistently to the tasks presented she might have scored a few points higher," Scally admitted. "However, my impression from these tests," she continued, "is that she will never develop intellectually beyond the level of a high grade moron."[66]

How could those constructing the narrative of Ann as "normal" counter such damning scientific evidence? Needing a medical expert to endorse their claim, they found psychiatrist Lawrence Collins, senior resident physician at the New Jersey State Hospital for the Insane. On January 8, while lawyers for Mrs. Maryon Hewitt McCarter declared they would produce expert testimony proving that Ann was feebleminded, Collins certified that Ann was "mentally sound."

In his statement, Collins declared that after an examination of Ann he found her to be "perfectly normal in every respect." He based his conclusion not on mental test scores but on her social abilities. He found her widely read (familiar with Shakespeare and Dickens), fluent in French and Italian, and possessing a wide range of interests. If she lacked anything, he argued, it was due to an "unwholesome environment" rather than any "pathological defect."[67] As eugenicists were beginning to argue in the 1930s, an inadequate environment—not just a pathological defect—

could and did damage a child. Yet Collins used this evidence not to justify sterilization of the child but to condemn an allegedly incapable mother.

Collins's analysis, along with the testimony of Ann and her three nurses, was enough to convince municipal court Judge Sylvain Lazarus that criminal charges should be filed against Ann's mother and the two surgeons. After hearing Ann testify in front of hundreds of curious onlookers, Lazarus addressed her with sympathy and respect, apparently persuaded by her story. For example, when Ann described Scally's testing procedures, she reported that she had asked Scally in disgust, "Why are you asking me these asinine questions?" Lazarus turned to her with a smile, duly impressed, and asked, "Did you really say asinine?"[68] When she mentioned she had corrected Scally's French pronunciation during another part of her questioning, the judge remarked, "Never dispute a psychiatrist. She probably marked you down a couple of points for that."[69] In his closing, he assured Ann, "You have been a wonderful witness. I know very few who think as clearly as you do."[70]

The discrepancy between Collins's and Scally's psychological evaluations led Lazarus to question the validity of mental testing. He asked Scally to give him a mental exam. She concluded, based on his answers, that he had a mental age of twelve. Lazarus facetiously responded, "I thought it was nearer eight."[71] The courtroom, "crowded to capacity," burst out laughing, while the judge concluded that the whole thing was "too silly for words."[72] In an apparent blow to the reputation of psychiatric testing, he dismissed the procedure as useless, calling into question the idea that intelligence was an easily measured genetic trait. The real issue, he believed, was not whether Ann was feebleminded but whether mothers should have the power to determine their children's fertility. "The necessity and the desire to bear children is something not idly to be interfered with," he warned.[73] He signed a warrant for the arrest of Ann's mother and the two surgeons, satisfied that mayhem had been committed by the three.

Had the story ended here, it would have been a hindrance rather than a help to the eugenics movement. As the media emphasized, Judge Lazarus believed that Ann's individual rights had been violated. But in the 1930s concern over individual rights had to compete with concern for the health and stability of the country as a collective body. In a time of crisis, personal rights had to be sacrificed for the common good—and eugenicists convinced many that the 1930s was such a time. They succeeded by emphasizing that eugenic sterilization was a "better deal for the child"—that all children should have the birthright of a sound mind in a sound body—and that such sterilization would ultimately preserve and strengthen the American family as a social institution.

In the only historical analysis of the Cooper Hewitt case, Daniel Kevles argues in a footnote that it "may have raised the general public's sensitivity to the misuses of sterilization."[74] But he does not discuss the powerful counter-narrative in the case, which was influenced by the updated campaign strategies of eugenicists in the 1930s. Supporters of Mrs. Hewitt reversed the charges, suggesting that it was Ann, not her mother, who suffered severe abnormalities. Ultimately, this narrative held greater sway with the presiding judge, as well as with public opinion.

Mrs. Hewitt, as her lawyer insisted on calling her, presented herself as a concerned mother. She declared that her treatment of her daughter had been "only what a kind mother would do for a daughter to whom she has been devoted."[75] From the moment of Ann's birth, her mother testified, she struggled unsuccessfully to raise her as a normal child. Mrs. Hewitt's close friend Dr. I. L. Hill testified that Ann's birth was premature (she weighed only 3.5 pounds), and, as a result, her physical and mental development was slow. Because premature babies usually survive only with careful nursing, he argued that Ann's survival was itself proof of Mrs. Hewitt's "exceptional motherly care."[76] She tried to give her a good education and sent her to the "best of schools," Maryon explained, but, unfortunately, Ann was not interested in furthering her education and had to be withdrawn.

Though Ann's lawyer testified that the daughter lived like a "poorhouse waif," Mrs. Hewitt argued that she had lavished expensive clothing and gifts on Ann. Embellishing this portrait, Ann's stepbrother came forward to defend Mrs. Hewitt. He declared that she had bought Ann forty pairs of shoes, one dozen evening gowns, and thirty hats. In addition, he claimed that twice she had made trips in "special railroad cars to insure privacy for her [Ann's] pet dog."[77] His testimony was supported by Helen Johnston, a buyer for the New York firm of J. M. Gidding & Co., who claimed that Mrs. Hewitt spent $35,000 a year on clothing for her daughter. Her purchases included expensive furs ("coats of ermine and chinchilla"), imported lingerie, and the "most expensive articles" in children's dresses.[78]

As evidence of her hardship in raising an obviously "backward" child, Mrs. Hewitt presented records of doctors' reports dating from a trip to the south of France in 1924. Two different doctors determined after examination that Ann suffered from "retarded growth" and "arrested mental development," which had led to "instability of mind and impulsive tendencies."[79] Along with Scally's conclusion that Ann's mind would never develop beyond that of a "high grade moron," Mrs. Hewitt used these medical reports to prove that she had Ann sterilized not out of personal greed but out of maternal concern. It was not that she did not want Ann to have children, she explained, but that she feared her mental condition would lead her into "moral difficulties."[80]

The operation, her lawyer explained, was performed both for Ann's own protection and for "society's sake." Ann's "erotic tendencies" had led her into a great deal of trouble. In 1932, she had become "infatuated" with a chauffeur and planned to run away with him. Her mother found love letters written by Ann that "justified their immediate destruction" and paid thousands of dollars to ensure that they were destroyed. One particularly ardent letter contained locks of hair from both her head and her pubic region (a fact Mrs. Hewitt disclosed only to the surgeons when requesting the operation; she refused to include this sexually explicit evidence in her affidavit). In addition, Mrs. Hewitt believed her daughter to be "easily infatuated by men in uniform," addicted to masturbation, and "familiar" with a hotel bellboy and a "negro porter on a train."[81] Ann's mother had spent years monitoring her daughter's actions, trying to block Ann's infatuations from developing into something more serious.[82] In this narrative, Ann, and not her mother, posed a threat to family and society; based on her actions, eugenicists believed she would not make a "desirable mother."[83]

Clearly Mrs. Hewitt's distress at her daughter's actions stemmed from the fact that Ann threatened her mother's rigid boundaries of class and race, as well as propriety. Ann challenged her mother's authority by independently asserting herself in relationships with men outside her own class and race. In the 1930s, such youthful rebellion was not new or uncommon. Scholarship in the history of gender and sexuality documents the way the behavior of youth, particularly female sexual behavior, became a challenge to the traditional social order in the 1920s.[84]

What *was* new in the 1930s and clearly articulated in the Cooper Hewitt case was the use of eugenic sterilization as a family-centered solution to the problem of female sexuality. Sterilization had become the private decision of a mother who sought control of her child's habits. Completely divorced from its institutional setting, sterilization offered Mrs. Hewitt a way of enforcing a standard of reproductive morality in her daughter. In the process, it raised the stakes of combining eugenic ideology with reproductive technology. Without the barriers of institutional commitment or professional

certification, sterilization could become a private family matter. If used con-
scientiously, as sterilization advocates hoped it would be, this family model
could have a far greater effect on improving civilization than any public institutional model would. As Gosney commented in reference to the Hewitt case, "Voluntary sterilization in private practice is of great value. Many
students of the subject are convinced that during the next generation the
great bulk of sterilizations in the United States will be of the voluntary [that
is, extra-institutional] type."[85] Their assumption was correct; beginning in
the 1960s, hundreds of thousands of women (primarily black, Native American, and Puerto Rican) were "voluntarily" sterilized in private hospitals,
often allegedly by doctors and social workers offended by their fertility
patterns.[86]

Mrs. Hewitt's case had the full support of the surgeons who had
agreed to perform the operation. Tillman justified Mrs. Hewitt's actions
as those of a caring mother, claiming he would have done the same thing
"had it been my own daughter." The doctor, who was also a member of
the San Francisco Lunacy Commission, acknowledged that Ann's case
was unusual in that she did not reside in an institution for the feebleminded, as the law specified. But the situation was virtually the same, he
argued. "Where a family has sufficient means, the feebleminded person is
taken care of privately." Ann had escaped the asylum only because her
mother could afford to care for her at home. But she was just as solid a
case for sterilization as any. "I feel justified both from a moral and scientific standpoint," Tillman continued. "It is an injustice to all concerned
to permit the feeble minded to bring children into the world."[87]

Boyd, who assisted Tillman in the procedure, also supported Mrs. Hewitt's request. He noted in his surgeon's report that Ann had "never developed beyond eleven years."[88] Surprised at the accusation of mayhem, he
remarked that he had not worried about the legal aspects of the matter,
"figuring a mother had the right to request such an operation, since the girl
then was a minor."[89] Though both doctors realized the California sterilization law did not explicitly give mothers the authority to have their
daughters sterilized, they presumed it to be a "moral matter"; as Tillman
said, "A mother is supposed to be acting for the best interest of her child."[90]

Tillman and Boyd were not the only ones confused about the boundaries of the law. Without legal precedent, the private sterilization based on
a mother's consent (and therefore considered voluntary) raised widespread speculation about the purpose and limits of the California sterilization law. Assistant District Attorney August Fourtner, prosecutor in
the Cooper Hewitt case, believed sterilization in private practice to be

illegal, a belief that supported the claim that the doctors had committed mayhem. The doctors' defense attorney, I. M. Golden, argued the reverse.

In pursuit of an effective argument, Golden contacted Popenoe at the HBF for advice on the legal aspects of Ann's case and asked him to testify as an expert witness. "In your opinion," he asked Popenoe, "was it proper to sterilize her by removing the Fallopian tubes, as a matter of medical and scientific procedure?"[91] After consulting Butler at the Sonoma State Home, Popenoe responded, "I suppose we should all answer negatively the question whether a young woman such as you describe would be a desirable mother." Significantly, Popenoe was not concerned about whether Ann's reported abnormalities were hereditary; though this was his specialty, he did not believe it was "particularly the issue in this case."[92] Instead, he based his answer on her suitability as a mother.

Popenoe's comments reflect the changing strategies of eugenicists in the 1930s. In 1935, the American Eugenics Society recommended that sterilization be used "even in cases where 'there is no certainty that the traits of the parents will be passed on to their children through heredity.'"[93] The goal of eugenics, the Society's president, Ellsworth Huntington, argued, was to improve society at its source: the home.[94] The way to achieve this goal was to ensure that the largest possible percentage of children were born in homes "best fitted to develop their character and intelligence" while the smallest percentage were born in homes where "parents are unable or unwilling to provide good training both intellectually and morally." The most important factor, he explained, was the "character and vitality of the parents."[95] Under these standards, in Popenoe's opinion, Ann did not make the cut. For this reason alone, he (along with surgeons Tillman and Boyd and, ultimately, the deciding Superior Court judge) found the sterilization to be justified.

After extensive correspondence with Popenoe, Golden appeared confident that he could argue for the surgeons' innocence based on the eugenic-sterilization law. Like eugenicists in the 1930s, he was also careful to couch such an argument in terms that a jury would find meaningful by embedding it in family issues. As he confided to Popenoe before the case went to trial in August, "What I am going to try to prove is that, according to experience it is well, both for Society, for the individual and for the family group, that morons be sterilized."[96]

In his preparation, Golden probably studied a significant article published in the journal of the American Bar Association entitled "Liability of Physicians for Sterilization Operations." Authors Justin Miller and

Gordon Dean, noting the rising number of sterilizations performed both in public institutions and in private practice, sought to determine the civil and criminal liability of physicians who performed them. A central question in their article, as in the Cooper Hewitt case, was whether doctors who performed sterilizations were guilty of mayhem. The crime of mayhem consisted of the "unlawful and malicious removal of a member of a human being or the disabling or disfiguring thereof or rendering it useless."[97] When applying this definition to sterilization in the 1930s, legal analysts struggled to determine the primary function of the female sex organs. If reproduction was the primary function, then mayhem indeed could be applied. But if it was the gratification of sexual desires, then sterilization did not render the organs useless.

The debate over the primary function of female sexuality—procreation or pleasure—was a result of the changing sexual morality of the 1920s. As women challenged the Victorian ideal of passionlessness and modern sexologists declared that sexual desire was a natural and normal characteristic of women as well as men, pleasure replaced procreation as the universal biological function of sex. Pleasure became a universal right, while procreation became an exclusive privilege. Thus, charging doctors who sterilized in private practice with mayhem rarely resulted in a conviction in the 1930s. The outcome depended on the interpretation of the phrase "rendering it useless" in individual cases. Another legal analyst argued shortly after the Cooper Hewitt case that it also depended on the "attitude of the courts concerning the morality of sterilization." Hartley Peart, a lawyer, noted in 1941 that "until recent years the attitude of all civic bodies, as well as the courts, was that sterilization was immoral in all aspects." But thanks to the work of advocates such as Dickinson and Popenoe, "in California, and in many other states, public opinion has progressed to the point of approving sterilization for the purpose of rendering the unfit incapable of procreation."[98] Such approval was confirmed in the Cooper Hewitt case, where the charge of mayhem for sterilization was defeated.

After months of collecting evidence to sway the jury, defense attorney Golden and prosecuting attorney Fourtner came face to face in the courtroom of Superior Court Judge Tuttle to try the surgeons on August 14, 1936. They selected a jury of twelve individuals whose task was to determine whether the surgeons were justified in "depriving Ann Cooper Hewitt of the right to motherhood on medical, eugenic, or other compelling grounds." The state intended to show that the surgeons were not justified but rather acted "maliciously and feloniously." The defense planned to prove that the doctors acted "ethically and in good faith"—

that Ann "had to be kept from motherhood" because of her mental and physical "unfitness."[99] Both intended to appeal to the jury by focusing on the family as a central theme in the case.

But Judge Tuttle never gave the jury a chance to make a decision, for after less than a week he dismissed the case as a "useless expenditure of public funds." He believed that the prosecution, who had relied on the testimony of Ann, her attendants, and the surgeons, had failed to make a case against the defendants. Most significantly for the future of sterilization practice, Tuttle declared that sterilization was not a crime in California; he thus established a precedent, as this was the "first ruling on the right of doctors to sterilize minor persons at the request of the parents or guardians."[100] Though the prosecution attempted to appeal the dismissal, the State Supreme Court declined to reopen the case just one week later.

Sterilization advocates, relieved at the decision, recognized its significance. The widespread coverage, the focus on the family, and the judge's dismissal all served to popularize sterilization as a moral solution to the problem of female sexuality. Butler collected copies of the judge's decision to send to private physicians "who may wish to consider the operation on private individuals in their respective communities."[101] Popenoe, who had advised Golden on the case, argued that regardless of whether Ann suffered from a hereditary deficiency, sterilization was justified by her sexual transgressions, which indicated she would never make a "desirable" mother.[102] Excited by the judge's decision, which validated this argument, Popenoe wrote to Golden, "May I join my congratulations to the flood of them which I am sure you have been receiving for your smashing victory. . . . This case will be a milestone in the judicial history of sterilization in California."[103] Sterilization advocates were no longer required to provide proof of hereditary deficiency; questionable sexual behavior was justification enough for sterilization. The determining question was no longer, Will she spread her genetic defect to her children? It was, Will she make a desirable mother? This change paved the way for widespread use of sterilization as a way to regulate motherhood.

Though today the case has been all but forgotten, the battle between Ann Cooper Hewitt and her mother familiarized Depression-era Americans with sterilization as a social remedy. One popular magazine noted that the suit "evoked public interest in sterilization." Before the case, few realized that twenty-three thousand Americans had been sterilized, ten thousand of them in California.[104] Yet even after widespread coverage of the trial, no public uprising or protests against sterilization occurred. Convinced that in a time of crisis personal liberty should be sacrificed for the common good, many spoke in favor of it. For example, in 1937, just one year after the Hewitt case, *Fortune* surveyed its readers on the issue and found that 66 percent favored compulsory sterilization of "mental defectives," while only 15 percent were opposed.[105]

Another indication of the increasing public interest in sterilization in the 1930s comes from Hogue's weekly column, "Social Eugenics," in the *Los Angeles Times Sunday Magazine.* The *Times,* at the suggestion of Gosney (president of the HBF), established the column in 1935 and continued it through 1940. Hogue introduced readers to a number of eugenic issues, stressing the necessity of eugenic sterilization in order to advance modern civilization, warning of race suicide, and updating readers on developments in the Cooper Hewitt case.

Hogue intended his column to evoke public interest in eugenics. "To appeal to the masses," he acknowledged, "one must speak in the language of the masses." As his column generated massive correspondence, he noted encouragingly the rise in public support for sterilization. "I am learning more about the actual reaction of the average person to the formulas of modern disciples of social eugenics from my correspondence than from all the books and lectures I have read on this subject."[106] In a column written four months after the Cooper Hewitt verdict, Hogue noted:

In casting over my correspondence on Social Eugenics . . . I find there is a growing interest in a more extended enforcement of State sterilization laws. This increasing interest comes chiefly from women, most of whom are mothers. Many letters come from women who confide to me in confidence conditions of families in their immediate neighborhood. They see families of subnormals, both mentally and physically, increasing without let or hindrance and they write me that many of the mothers would be willing to submit to a surgical operation which would make sure that they would give birth to no more children mentally unequipped to care for themselves in the modern struggle for existence.[107]

His correspondents expressed concern over women whom they believed ought not to have children. More significantly, they suggested that these mothers themselves agreed. Coupled with the letters Gosney and Popenoe received from women sterilized in California institutions, these letters reveal the existence of a widespread belief in reproductive morality in American society. "Reading your column," a thirty-year-old woman wrote, "I am amazed in this day and age right-thinking people let imbeciles have children. Why, and when will a law be passed to stop this?"[108]

The concept of reproductive morality was beginning to take hold in the 1930s. Public sentiment in regard to sterilization changed dramatically during this period; sterilization symbolized not mayhem but morality. Its promotion and use on the "unfit" illustrated the benefits of inculcating reproductive morality, as a form of social responsibility, to improve American society. As the California Appellate Court concluded in its decision in another 1936 sterilization case, *People v. Blankenship*, "If reproduction is desirable to the end that the race shall continue, it is clearly desirable that the race shall be a healthy race, and not one whose members are afflicted by a loathsome and debilitating disease."[109] Sexual pleasure was an individual right; reproduction was a collective issue whose goal was the betterment of the human race.

The 1930 White House Conference on Child Health, the Cooper Hewitt case, and the *Los Angeles Times*'s "social eugenics" column all reinforced the eugenic argument that Depression-era poverty was not the cause but only a symptom of a more deeply rooted family pathology. Even the wealthy Hewitts could not survive intact because both mother and daughter abandoned their culturally prescribed domestic roles. Their social and sexual transgressions reminded readers of the devastating cost of the individualism and self-gratification of the 1920s, which eugenicists believed had weakened the bonds of the American family by luring women out of the home.

Instead of experiencing a dramatic decline in influence during the heredity/environment controversy in the 1930s, as previous historians have argued, eugenicists used the new emphasis on environment to their advantage. If the home environment shaped a child's personality and development, then it was all the more important, eugenicists concluded, to scrutinize the home. "Parents produce faulty children by bad rearing as well as by bad heredity," Roswell Johnson argued while promoting an extension of the California sterilization law to authorize the operation on noninstitutionalized women at public expense.[110] Claiming that motherhood should be selective and fertility should be an option only for those who proved "fit," eugenicists raised the stakes of their campaign. By recognizing the influence of the home environment, some eugenicists also shifted their focus toward positive eugenics beginning in the 1930s. In particular, Dickinson, Terman, and Popenoe began to focus on marriage and family counseling as a legitimate eugenic strategy to counteract the declining birthrate by promoting the procreation of the white middle class.

CHAPTER FIVE

"Marriage Is Not Complete without Children"

Positive Eugenics, 1930–1960

The story of Ann Cooper Hewitt, which riveted the American public in 1936, drew attention to the problems of gender and sexual morality in modern society and to eugenics as a potential solution. Her sexual rebelliousness convinced many that she would not make a desirable mother, and thus she needed to be sterilized. But eugenicists believed that more than sterilization of the unfit was needed to bolster the American family and to produce a healthy and virtuous citizenry. With motherhood and family preservation at the center of their campaign beginning in the 1930s, many turned to positive eugenics, or promoting the procreation of the eugenically fit, as an additional strategy. By educating young adults about the principles of selecting a eugenically fit mate and counseling those already married on how to preserve and protect their marriage and family, eugenicists hoped to alter the way ordinary Americans thought not just about Cooper Hewitt but about their own decisions regarding marriage and procreation. Increasingly anxious about new challenges to the sanctity of the family—namely, changing gender roles and the increasing visibility of homosexuality—they directed their attention to negotiating marriage and parenthood.

From the thirties through the fifties, eugenicists promoted family stability as an essential component of modern progress and civilization. Children "sound in mind and body" could be born and raised only in the homes of happily married couples. The rising divorce rate—one in six marriages ended in divorce in the 1930s, one in four by 1946—as well as a declining birthrate suggested that the American family was becoming increasingly vulnerable.[1] Indeed, the Depression and war severely disrupted family life. During the Depression, rates of desertion soared so that by 1940 over 1.5 million married women lived apart from their husbands. During World War II, four to five million "war widows" suffered prolonged separation from their husbands. Yet many believed that a stable family was of even greater necessity during the war than before. As Linda Gordon points out, "If good morale on the home front was important to the war effort, it seemed that family stability should provide that morale."[2]

Economic necessity in the 1930s and wartime labor shortages in the 1940s also influenced marriage and family patterns, as the number of married women working outside the home dramatically increased. During the 1930s, the percentage of wage-earning women who were married rose from 28 percent to 35 percent.[3] Between 1940 and 1944, five million women entered the workforce (raising the total of working women to nineteen million). Of these, the number who were married with young children rose 76 percent, to 1.47 million.[4]

As more women found themselves separated, deserted, or divorced during these two decades, as well as in the labor force, they implicitly challenged the assumption that their primary purpose was in the home. To eugenicists such as Robert Dickinson, Lewis Terman, and Paul Popenoe, these trends explained the declining birthrate and underscored the importance of stable marriages and families to prevent race suicide. As a result, they shifted their focus, beginning in 1930, from preventing procreation of the unfit to promoting the marital and family stability of the white middle class.[5] In particular, they reached a widespread middle-class audience through psychological testing and "marital-adjustment" counseling, which were promoted in popular books, magazines, television programs, and marriage-counseling centers.

The positive eugenics campaign of the 1930s and 1940s paved the way for the pronatalism of the 1950s. Between 1940 and 1957, the fertility rate in America rose 50 percent.[6] Marriage rates soared to an all-time high as well. At the close of the 1950s, 70 percent of all women were married by the age of twenty-four, and the median age of marriage for women was twenty. Scholars have questioned why postwar Americans turned to marriage and parenthood with such enthusiasm and commitment. Many argue that the trend was an inevitable result of a return to peace and prosperity after two decades of strain on family life. They suggest that the opportunity to have a family filled a deep emotional need. Others link it to anxiety about the Cold War.[7] No one, however, has considered the role of eugenics in creating a powerful pronatalist climate in postwar America. Since scholars of eugenics portray the movement as weak and discredited after 1930, historians have missed an essential link between eugenic ideology and postwar culture.[8] As eugenicists broadened the definition of eugenics in the 1930s to include environmental and cultural influences on human development, they strengthened their campaign. Not only a mother's genes but also her competence to raise a child in a proper home environment became a eugenic concern. This redefinition allowed eugenicists to pursue marriage and family counseling as an essential strategy for strengthening the race.

Reproductive Morality

Through the first half of the twentieth century, eugenicists targeted female social, sexual, and reproductive behavior as a eugenic threat. While the "new woman" chose education, a career, and birth control over marriage and motherhood, "moron girls"—the label for working-class, sexually suspect, institutionalized women—were "extremely prolific." Both behavior tendencies marked the complete repudiation of Victorian morality. As promiscuity replaced purity, and sexual recreation replaced procreation, eugenicists frantically sought a modern construction of morality that would keep society from declining into sexual excess. "We have given birth to a generation in which anarchy in faith has overthrown authority," *Los Angeles Times* columnist Fred Hogue declared in 1936. "Out of the present disorder and unrest a new order must arrive if civilization is to be preserved."[9] As marriage was no longer a prerequisite for sexual intercourse, eugenicists needed to ensure that children were born to married parents and, ideally, were raised by full-time mothers. They therefore introduced and promoted what I have called "reproductive morality"—a eugenically based ideal that called on prospective parents to consider their progeny's potential impact on the race. Well-known sexologist and eugenicist Havelock Ellis articulated the goals of reproductive morality by arguing that potential parents should always consider whether their offspring would improve the quality of future generations. If a couple were not fit "to be the fine parents of a fine race," then in his opinion they should not marry.[10] Especially effective during times of crisis, such as depression and war, the emphasis on reproductive morality reminded Americans that parenthood was a privilege that had implications not just for individual families but for civilization as a whole. For example, one female doctor wrote in 1950 that she and her husband had produced only two children. "Now I know we should have had six," she declared, "because both our children are old enough to show high IQs. I am appalled to think how we have denied the human race four more of the leadership type."[11] Twentieth-century reproductive morality promoted marriage and motherhood as a central goal of womanhood. Though women had won citizenship through suffrage and had gained economic and educational opportunities, eugenicists suggested that none of these achievements compared with the sense of fulfillment that marriage and motherhood would bring.[12]

This sentiment was best exemplified in *Janet March,* a popular coming-of-age novel written by Floyd Dell, a cultural radical and well-known novelist living in Greenwich Village during the 1920s and 1930s. The story centers on the struggles of young Janet to find a meaningful place for herself in the 1920s. Living in a "crazy world," in which female sexuality had taken on new and significant meaning, Janet felt lost, confessing, "I don't know just exactly what I am."[13] Coming of age in a decade when the "incidence of premarital sexuality jumped sharply, to nearly fifty percent of the cohort," Janet was shocked to discover that some of her friends did not stop at petting (92 percent of college coeds engaged in petting during the 1920s) but were engaged in sexual activity.[14] Janet herself felt torn about premarital sex. Her friends seemed to enjoy it. But she also learned from her cousin of the problems of institutionalized "wayward girls" who had borne children out of wedlock. Harriet explained that the infants had to be taken away from these "ignorant and restless mothers" who were not capable of raising children and who "ought not to have these babies at all."[15]

Dell purposely juxtaposed these examples of modern sexuality to illustrate both the temptation and the price of sexual freedom to women in the 1920s, issues he himself grappled with while living in Greenwich Village. Searching for excitement, unsure of who she was or what she wanted, Janet hoped to find herself in the experience of sexuality. Though she is not in love, she has sex with a man at the age of twenty-one, declaring, "I have a right to this experience." Unfortunately, she becomes pregnant, and cousin Harriet takes her to a female doctor in Chicago for an abortion. Though the doctor assures Janet that she is "a magnificent specimen of a woman" who will have "fine babies," Janet cannot but compare her predicament to that of the wayward girls. She vows to change, remarking, "There must be some *real* use for me—I *mustn't* just go to waste."[16]

Finally, after 250 additional pages of soul-searching, Janet finds her mate.

Discovering that she is pregnant once again, she fears the ruin of her relationship. When Roger instead embraces the news, Janet realizes that she has found her purpose in life. "All my life I've wanted to do something with myself," Janet declares. "Something exciting. And this is one thing I can do. I can help create a breed of fierce and athletic girls, new artists, and singers."[17] Smart, determined, and willful, Janet realizes that, even with the opportunities in the 1920s to explore sexual freedom and a career, she would gain the greatest sense of fulfillment from rearing children.

Dell's message is clear. If modern marriage and motherhood were promoted as happy and fulfilling modern ventures for women rather than as forms of servitude (as often touted by contemporary feminists), then women would be more readily drawn back into the home. One reviewer, who found the novel "amazing," characterized Janet March as "the most modern of all modern heroines."[18] Though other reviewers recognized that the story of *Janet March* did not present a new vision of womanhood but rather a recycled tale of "unquestionable orthodoxy," its vision of companionate marriage and prolific motherhood would retain authority through the 1950s.[19]

Janet's search for fulfillment in marriage as well as motherhood underscored the shift, beginning in the 1920s, toward the ideal of a companionate marriage based on friendship and sexual satisfaction. "I used to think that [having babies] was the only reason for getting married," Janet remarks, "but now I feel that doing things together is quite as important as having babies."[20] The term *companionate marriage* was coined in Judge Ben Lindsey's 1927 guide to marital happiness. Responding to feminist demands for sexual liberation, marital advice givers "sought harmony between the sexes by reforming what seemed the most oppressive elements of Victorian marriage." Educators, social workers, psychologists, sociologists, physicians, and eugenicists wrote books and articles on sex and marriage in the hope of channeling female sexuality into marriage. In this way, they sought to redefine the institution in egalitarian terms. Ernest Groves, founder of the field of marriage counseling in the United States, noted in 1928 that sexual expression had become increasingly accessible outside of marriage and therefore "a revision of norms concerning marriage was imperative in order to protect the institution from decay."[21]

In their analyses of companionate marriage, both Nancy Cott and Christina Simmons argue that the "decline of organized feminism after 1920" allowed this conservative, heterosexual ideal to attain cultural hegemony by the 1930s.[22] Neither addresses the role of eugenics—a movement supported by many feminists—in bolstering the companionate-marriage ideal. Beginning in the 1930s, eugenicists expanded their campaign by promoting marital and maternal happiness as a significant and fulfilling goal for modern womanhood.

But they limited their campaign to the middle class. Certain women, of course, were not meant for motherhood. Revealing the darker, more manipulative side of working-class female sexuality was the popular novel *Stella Dallas,* written by Olive Higgins Prouty and adapted for Hollywood by King Vidor in 1937. Stella was an uncouth, lower-class woman, desperate to escape her life in Cataract Village. Flirtatious and blessed with "stacks of style," Stella, using her charms to work "their blinding enchantment," convinces the classy lawyer Stephen Dallas to marry her.[23]

Dallas soon realizes he has made a terrible mistake. Aware of her limitations and "little crudities," he initially tries to "rub down her rough edges." But within a few months he notes that "it was going to be as difficult to stamp out Stella's vulgarity as to rid a lawn of the persistent dandelion once it gets its roots down." She continues to flirt with other men and seems more concerned about cultivating her appearance than about cultivating their relationship. "I must be patient," Stephen told himself. "It is only that she has been bred differently." It seemed "instinctive" for Stella to flirt with other men.[24]

There is little hope for the marriage because Stella is incapable of changing her "little crudities." While she does not suffer from any hereditary flaws, she cannot escape the environment in which she grew up. In the 1937 movie, Barbara Stanwyck gave an Academy Award–nominated performance as Stella Dallas, making her character increasingly crude and tasteless as the drama unfolds.

A baby arrives during their first year of marriage, but motherhood does nothing to tame Stella. Neighborhood women find the child, Laurel, a "nice little thing" but wonder how long she will stay that way with Stella as a mother. "Such a woman doesn't deserve to have a child," they declare. Stephen, fed up with his embarrassing wife, moves to New York and ends the marriage.[25]

In New York, he reunites with his high school sweetheart, a beautiful, refined, and tasteful widow and mother of two. Stephen falls in love with Helen Morrison and invites his now-teenaged daughter Laurel to spend time with them in Helen's Long Island mansion. Helen immediately takes to Laurel as if she were her long-lost daughter. Stella, realizing that Laurel will have a better life with Helen and ex-husband Stephen than with her, gives up her maternal rights to her daughter. Explaining her decision to give Laurel to Stephen, Stella notes, "Laurel takes after her father. I can't see a trace of me in Laurel. Nobody can. She's so refined, and sort of elegant in her ways.... Why, if that girl didn't have *me* shackled round one foot everywhere she goes, she'd just *soar.*" The story ends happily for everyone except Stella, who is left alone; Laurel fits into New York's high society and marries into the right family.[26]

Stella Dallas reminded readers and moviegoers that motherhood was not for everyone; it was a class-based ideal. Though Stella is not "feeble-minded," she cannot escape the vulgarities of her working-class upbringing. Recognizing that her daughter belongs with her ex-husband and his new wife, she unselfishly gives up her maternal claims to Laurel. Her self-sacrifice allows her daughter to have a better life. Stella recognizes the importance of reproductive morality, relinquishing maternal rights for the betterment of society.

Eugenicists, believing that women like Stella Dallas would not make "desirable" mothers, changed strategies in the late 1920s and early 1930s, shifting their emphasis from heredity to environment. This shift allowed them to target motherhood as an essential role in raising eugenically fit children. For negative eugenics, sterilization could be used to weed out undesirable mothers, regardless of whether they carried any hereditary deficiency. But the shift also allowed eugenicists to expand their role as educators for middle-class marriage and motherhood in the field of positive eugenics. If "the great middle class of our population and the upper class in general are not having enough children to replace themselves," as American Eugenics Society (AES) president Ellsworth Huntington argued in 1935, then they needed to be persuaded to increase their progeny.[27]

Eugenicists unleashed an educational campaign in the 1930s to enlighten white middle-class Americans about their eugenic duty. "Eugenical education builds the standards of the people," argued Harry Laughlin; "it inspires racial, nation and family-stock loyalty."[28] After forming the Southern California branch of the AES, Popenoe declared, "It is evident that education is now the most promising field for the practical and constructive development of the positive side of eugenics."[29] He therefore devoted the branch to outreach and education. It sponsored lectures on topics such as "proper safeguarding of the marital relationship," "educating high school boys for marriage," and "can romance be reconciled with eugenics?" (445 people attended this lecture in 1932 in the Los Angeles public library).[30] The Society asked junior high and high school teachers to report on the application of eugenics in the public school curriculum; it noted that educational leaders recognized the "social value of a general knowledge of eugenic principles" and that eugenics already had a place in public school curricula.[31]

But, according to Popenoe, not enough was being done at the college level. He argued in 1930 that "all educational institutions should place [marriage and parenthood] second to no other objectives" because of their importance "personally, socially, and racially." Given the low marriage rate and birthrate of men and women college graduates, clearly such training was lacking. Popenoe suggested that college faculty should be selected based on their moral as well as their educational credentials; unmarried women "should be supplanted as instructors by those who have successfully brought up families of their own." In addition, the college curriculum should be structured around marriage and family and should include courses on human heredity, which were necessary for proper selection of a mate. But the largest current problem, Popenoe stressed, was that students no longer understood the central place of the family in society. They should be taught to recognize how important it is, "social and racially, to provide a citizenship that will work and vote intelligently for the conservation of the family." Nothing less than civilization as a whole was at stake, he argued. "History shows that, in general, the highest state of civilization has accompanied the strongest and most vigorous family life." What colleges in modern America needed was not "new professors, new laboratories, new endowments; it is merely the need of recognition, on the part of all those who make up a college, that the family is the central fact in human life."[32]

Education campaigns extended beyond the classroom to other aspects of modern life. For example, the AES sponsored a conference on "recreation and the use of leisure time in relation to family life" in 1937. Eugenicists had come to recognize that "the eugenic ideal must become a part of every organized effort to improve human conditions." Simple, noncommercial recreation would offer youth an opportunity to establish normal relationships with the opposite sex, as well as give training in "character, in physique, and in enjoyment of life."[33] They praised the health program of the Boy Scouts as an ideal in need of wider application. The program taught boys how to manage their lives, interact with others in community life, and learn leadership skills: all qualities they believed to be necessary for a happy marriage and family. Though the YMCA and YWCA were beginning to offer joint programs, additional co-ed recreational activities were needed to condition young people "to the maintenance and increase of those standards of family life which are an essential part of the eugenic ideal."[34]

Recognizing that youth in the 1930s possessed the curious and adventurous nature that had gotten Janet March into trouble, eugenicists warned, "If we do not plan for the two sexes to mix in play they will arrange to get together under other circumstances anyway." If eugenicists could plan co-ed recreational activities that fostered healthy attitudes toward marriage and parenthood, it would "give to our young people a greater appreciation of discrimination in selecting their mate."[35] It would help to ensure that procreation would occur primarily in well-adjusted, happy marriages.

The birth-control movement also began a campaign to strengthen the American family at this time; this campaign paralleled the shift from negative to positive eugenics. Advocates emphasized the need to reinforce family values and argued that birth control would produce "wholesome family life." The name change of the national organization from the American Birth Control League to Planned Parenthood Federation of America (first proposed in 1938 and adopted in 1942) signified this new, "positive" birth-control program, which encouraged a "sound parenthood . . . as a major means by which this nation can be maintained strong and free." This new emphasis allowed many eugenicists who had formerly opposed birth control to support the cause. As Henry Fairchild, president of the AES, remarked in 1940, eugenics and birth control "have come to such a thorough understanding and have drawn so close together as to be almost indistinguishable."[36]

Robert Dickinson: The Doctor as Marriage Counselor

Robert Dickinson, who had helped to generate an understanding between the two movements by linking sterilization to birth control, also played a significant role in the positive-eugenics movement by launching his own marital-education campaign. He believed that sexual maladjustment in marriage was the primary reason for the rising divorce rate and lowered birthrate. Assuming that marriage "should be the ultimate in human relationships," he devoted most of his professional life after 1920 to the promotion of stable marriages through sexual adjustment, leading a *Reader's Digest* journalist to proclaim him a "pioneer in marriage counseling." In particular, he called on doctors to assume an increased role as "spiritual adviser" to patients, asking them to diagnose not just physical problems but moral and social ones as well.[37]

Central to his educational strategy was the idea of "preventive gynecology" as a way of ensuring marital adjustment by placing the gynecologist in the role of marriage counselor. As Dickinson argued in 1928, "Examination for fitness has become customary for all occupations save marriage and parenthood. Eventually common sense may be expected to demand a similar preparation before deciding on matters so important to the life of the individual and the race." He therefore proposed that every couple undergo examinations by a doctor before announcing their engagement. The examinations would include searching for any "defects" or diseases, a study of the woman's "genital and pelvic anatomy with reference to marriage and childbearing," and sex instruction. He believed that physical examination of the woman would prevent a number of potential problems that could cause divorce or adultery, including frigidity, abortion, and "unwise postponement of childbearing." The doctor's role was also to prepare the future wife for intercourse by easing any "association of pain or fear with the entry of the male," which sometimes included cutting a thick hymen ten days before the marriage. Believing that sexual maladjustments were often preventable, Dickinson urged the medical profession to recognize that its role included that of marriage counselor. In addition, he drafted chapters for a book (never published) entitled *The Doctor as Marriage Counselor,* which included sections on how to train and review the work of physicians as marriage counselors, as well as more general information on coitus, contraception, sterilization, and sterility.[38]

In 1931, Dickinson promoted his concept of preventive gynecology in *A Thousand Marriages: A Medical Study of Sex Adjustment* based on his forty-year study of marriages and 1,098 records. While half his patients had come to him initially for "problems of childbearing," Dickinson noted that most of these problems were preventable. Those women in satisfactory or at least "not-complaining marriages" were more fertile than those who were unhappy in their marriages, he argued, often because of sexual maladjustment. With counseling from a trained doctor, sexually maladjusted women could develop a more positive attitude toward coitus, resulting in better marriages and more children.[39]

In addition to misinformation and a lack of guidance, another factor threatened the sanctity of marriage and the family in modern America, according to Dickinson and other eugenicists: homosexuality. Though the category of homosexuality had received little attention from eugenicists, their newfound focus on family and motherhood, as well as the increasing visibility of urban gay and lesbian subcultures, caused them to address the issue in the 1930s. Most believed Havelock Ellis's claim that true "congenital inverts" were "powerless to change [their] inclination."[40]

Sexual inversion "connoted a complete reversal of one's sex role": male inverts were feminine in behavior and appearance, female inverts masculine. As the congenital invert was not likely to reproduce, she would not pass on her genetic aberration to offspring. Many other women and men, however, were latent homosexuals, easily influenced by true "inverts." With education and training in a heterosocial atmosphere, eugenicists believed these individuals could lead "normal" (that is, heterosexual) lives as husbands and wives.[41]

Homosexuality was thus constructed as a severe form of sexual "maladjustment," one which threatened to weaken marriage and family even further. Concerned about homosexuality's effects, Dickinson launched a six-year study beginning in 1935 through the Committee for the Study of Sex Variants (many members of which were eugenicists). The purpose of the study was to "assist physicians in identifying and treating individuals who suffered from 'sexual maladjustment' and to help prevent the spread of sex variance through the 'general population.'" The ultimate goal, of course, was to foster healthier marriages and more children by identifying and isolating those individuals whose sexuality threatened the "natural order" of heterosexuality contained within marriage. As Jennifer Terry argues, "The scientific making of the homosexual type was integrally connected to campaigns for encouraging hygienic heterosexuality among white people."[42]

Under the leadership of Dickinson, the Sex Variant study underscored the nineteenth-century theory that defined homosexuality as sexual inversion: "too much masculinity in women and too much femininity in men." As the study discovered, however, homosexuality was not always written on the body; male and female "inverts" did not always display physical characteristics of the opposite sex.[43] How, then, was it possible to detect homosexuality, ideally before too much damage had been done to the "normal population"?

Lewis Terman:
Sexual Maladjustment and the M-F Test

Just as mental testing emerged in the 1910s to weed out the "high-grade morons" who were passing as "normal" in the community, personality testing appeared in the 1930s as a way to weed out homosexuals. In both instances, Lewis Terman (also a researcher for the Sex

Variant study) played a starring role. In 1916, he had produced the first widely used individual measure of intelligence—the Stanford-Binet scale—which contributed to the increasing importance of mental and moral "normality" in twentieth-century American culture. In 1936, he published, with Catherine Miles, *Sex and Personality: Studies in Masculinity and Femininity,* which introduced the masculinity-femininity (M-F) test. The purpose of the test was "to make possible a quantitative analysis of the amount and direction of a subject's deviation from the mean of his or her sex in interests, attitudes, and thought trends."[44] By quantifying masculine and feminine traits, Terman hoped to standardize male and female behavior just as he had standardized intelligence, thereby providing a mechanism to weed out both mental and sexual "deviants." He believed that the M-F test, the first of its kind, would enable investigators to obtain a more "exact and meaningful, as well as more objective, rating of those aspects of personality in which the sexes tend to differ."[45]

The M-F scales included 456 items in seven different exercises and used standard personality-test procedures, such as Rorschach inkblots. They included multiple-choice questions, such as: "Marigold is a kind of fabric (+), flower (–), grain (–), stone (+)." A "+" answer gave the test taker a point for masculinity, a "–" gave a point for femininity (a significant choice of symbols: masculinity scored as positive; femininity, as negative). Also present on the test were yes or no questions, such as "Are you extremely careful about your manner of dress?" ("yes" signified feminine; "no" signified masculine), and true or false questions, such as "Children should be taught never to fight" ("true" was a feminine response, "false" a masculine one).[46] Test takers were not informed about the meaning of the symbols or the purpose of the test.

Though Terman and Miles did not venture to estimate whether these sex-based characteristics were determined by nature or nurture (or both), they still concluded from their study that men and women were fundamentally different. One reviewer claimed the book proved that "sex differences go deeper than the genitals."[47] Havelock Ellis found the book to be "a real achievement" and was delighted that gender differences could be scientifically quantified.[48] Yet, by quantifying such behavior and placing it on a spectrum with masculinity at one end and femininity at the other, Terman suggested that gender boundaries (or "boundaries between the sexes," as he would understand it) were easily transgressed. Just as the Stanford-Binet test raised anxiety that only a few degrees separated "feeblemindedness" from "normal" mentality (it would be easy, in other words, to "slip" into such a state), the M-F test provided evidence that only a simple deviation from masculinity or femininity separated the "healthy heterosexual" from the "homosexual invert." In this context, the sex variant became "a confusing border creature" between male and female, normal and abnormal, healthy and sick.[49]

As a result, many psychologists were eager to get their hands on the test, not only to measure the extent of their patients' behavioral deviance but to ensure that they themselves were sexually "well-adjusted." As firm believers in the power of personality tests to reveal an objective truth, mental-health professionals found that even they had something to gain from taking such tests. For example, Dr. Aaron Rosanoff, a Los Angeles–based psychiatrist, wrote Terman, "I am greatly interested in your masculinity-femininity test, for if it should prove to be a trustworthy measure of masculinity and femininity, it would be of great value in psychiatric practice, as we are daily confronted with the necessity of determining the psychosexual makeup of patients." Though Terman had sent him copies of the test to administer to patients, Rosanoff took the liberty of completing the test himself and was returning it to Terman for evaluation.

He then confessed his concerns:

> I am not homosexual, but over fifteen years ago Dr. William A. White of Washington undertook to analyze a childhood dream of mine, and at that time surprised me by offering it as his opinion that I had a concealed or repressed homosexual component in my makeup. If that is so, then I would state that I was never conscious of it, and have not become conscious of it since then, although I have made a candid search for it myself.
>
> However, according to psychoanalytic theory, it is not only possible for a person to have a homosexual component in his makeup without being aware of it, but the majority of cases where there is such a component are characterized by complete repression of same.
>
> I admit that this sounds like a joke, if it were not for the fact that I have on many occasions been able to fully satisfy myself that such a thing has happened in cases of patients of mine, and that the maladjustment which has brought them to me could be disposed of by revealing and ventilating this underlying mechanism.[50]

After looking at Rosanoff's test, Terman reassured him that "the opinion of your psychoanalyst friend is not borne out by your test scores."[51] In a scientific climate that viewed homosexuality as a pathological condition, anxiety about latent homosexuality could be relieved by the M-F test. Yet like an IQ score, the M-F score could take on inestimable power as a signifier of pathology: failure to acquire gender-appropriate identity, as revealed by an M-F score, suggested homosexuality.

Though Terman and Miles were not the first to suggest that homosexuals identified with the wrong sex (that is, that male homosexuals were "effeminate" and lesbians were "mannish"), their test provided evidence that gender-inappropriate behavior was the key to the "problem" of homosexuality. Yet there was a circularity in their argument. They validated the test by comparing test scores of "well-adjusted" individuals to those of homosexuals—thus already assuming that homosexuality signaled misplaced sex-role identification.[52] They crafted the test with the assumption that male homosexual interests and attitudes would be effeminate, while lesbian interests would be masculine. Thus if a man wrote that he would like any of the following careers (regardless of pay and assuming it was equally open to men and women)—chef, librarian, journalist, social worker, music teacher, clerk, singer, or artist—he would get a negative (feminine) score. (The "gender-appropriate" responses for men were architect, forest ranger, stock breeder, and soldier.)

It was not that Terman believed every male artist or aspiring male artist to be a homosexual. The test taker had 455 other questions on which to assert his virility. But, by assigning a positive or negative to every aspect of personality, Terman was reinforcing the polarization of masculine and feminine modes of behavior in early twentieth-century American society. Masculinity was the ideal; femininity its opposite. Thus, when searching for homosexuals to test, Terman requested "particularly effeminate, sissy" men, forty-six of whom he found at the U.S. Disciplinary Barracks at Alcatraz, and eleven in prison at San Quentin. Terman's men were selected by prison officers, who chose their subjects based on their effeminate behavior—not necessarily on their sexual orientation. Not surprisingly, these seventy-seven "passive homosexual" males tested "extremely effeminate—some of them more feminine even than the average woman, as measured by the scale."[53] Terman had rigged the test. He claimed his group of homosexuals was a representative sample, but it was selected on the basis of the very factor it was claiming to prove: effeminacy. The feminine appearance and behavior of his subjects confirmed their pathology as homosexuals:

The behavior of these male homosexuals, both as reported by them and as observed during the investigation, is as different from that of a group of average men as one could possibly imagine. It is well known that the average boy or young man makes every effort to keep from appearing effeminate. The passive male homosexual, on the contrary, takes advantage of every opportunity to make his behavior as much as possible like that of women. He not only accentuates any feminine qualities which he may possess, such as a high-pitched voice, but also attempts to imitate women in his speech, walk, and mannerisms. Practically every subject has adopted a "queen" name by which he is known among his associates. Their behavior often seems exaggerated and ridiculous, although in some cases the inversion of behavior is remarkably complete.[54]

Terman's "passive" subjects adopted not only a "queen" name but a feminine, subordinate role in both sexuality and society. The "active male homosexuals" whom he studied, however, generally scored in the masculine range. He concluded that they were "masculine in sexual responses, but the stimulus which provokes their sexual responses is a male instead of a female, though preferably a male of feminine personality." Active homosexuals, believed to assert the dominant sexual position of "penetrator," did not challenge the assumption that men should assume dominant roles in sexuality and society.

For lesbians, the reverse was true: "active" lesbians scored masculine on the M-F test and challenged the subordinate position of women in society, while passive lesbians generally scored feminine on the M-F test (thus demonstrating less maladjustment). In one case study, Terman noted that the passive lesbian was "small and of feminine build." She left home at a "tender age" because of an excessively brutal father. After college she married "an effeminate man who left her because he disliked sexual intercourse." Echoing Dickinson's argument that sexual maladjustment in marriage was a major cause of family instability, Terman concluded, "There is evidence that but for her unfortunate marriage she would have been entirely heterosexual." Her partner, however, assumed the role of a "devoted husband." She scored in the 75th percentile for college men on the M-F test, had a "boyish build," and dressed "rather severely," lacking the "feminine touch."[55]

Terman's objective was to create rigid standards of masculinity and femininity as a way of shoring up traditional gender roles and family patterns. The increasing visibility of gay men and lesbians—particularly what Terman believed to be "passive homosexuals" and "active lesbians"—suggested the potential fragility of heterosexual familial standards.[56] By either weeding out or "curing" homosexuality, Terman believed he could stabilize the American family and thereby strengthen the race.

Homosexuality could best be combated, he argued, with the use of the M-F test. "The problem of homosexuality attracts from scientists but a small fraction of the attention it deserves," he explained. "We are hopeful that the M-F test will prove a useful tool in this field of investigation. . . . For one thing, the use of the test will help to center attention on the developmental aspects of the abnormality, just as intelligence tests have done in the case of mental deficiency." Terman is, of course, referring to his own source of influence, the Stanford-Binet test, which he developed to standardize the concept of mental normality. "It is well known that the milder grades of mental deficiency can now be detected earlier than was possible a generation ago," he continued. "The same will in time be true of the potential homosexual. Early identification of the latter deviant is particularly to be desired, because we have so much reason to believe that defects of personality can be compensated for and to some extent corrected."[57]

Terman's belief that homosexuality was a pathology affected not only eugenicists and psychologists but also some homosexuals who wrote Terman after recognizing themselves in his study. A nineteen-year-old from Trenton, New Jersey, wrote Terman in 1939, "I am a homosexual and I don't know what to do about it." He could not turn to anyone local as he did not believe there was anyone who "knows anything about such a thing to give advice." Though he had known "for some time" that something was causing him "some nervous irritation," he did not realize its origin until reading Terman's works, in which he had identified himself as a homosexual. "Your writing gave me the explanation for my trouble," he explained. He experienced some of the circumstances Terman and other sexologists had labeled as influential factors: he lived in an "undesirable neighborhood" and had "few social contacts." Most important, his mother always took his side "and still does against my father."[58]

The young man had most likely read Terman's conclusions in his case studies of homosexual males (a chapter in *Sex and Personality*). Based on his study of "passive" homosexual subjects, Terman concluded that "the psycho-social formula for developing homosexuality in boys" included "too demonstrative affection from an excessively emotional mother" and "a father who is unsympathetic."[59] Recognizing his own parents in this formula, the young man determined he was a homosexual. "And now that I do know," he confided, "I'm quite upset." Desperate to adhere to Terman's standards of normality, he pleaded, "I want something to guide my future actions; something that will improve my present condition." In search of a solution to his condition, he begged, "Wouldn't you please give me a few simple suggestions I could follow?" Terman sympathized with his "problem" and suggested he contact a psychiatrist at Yale University.[60]

The M-F test might be disregarded today as merely bad science, but its significance should not be underestimated. Dozens of psychologists and psychiatrists wrote to Terman personally, expressing their interest in using the test on their patients. Many more purchased the test directly from McGraw-Hill. According to one historian, it was "perhaps the single most widely used inventory to determine the successful acquisition of gender identity in history and was still being used in some school districts into the 1960s."[61]

In addition, Terman's ability to standardize masculinity and femininity provided sociologists of the 1950s with key evidence that their culture was in danger of degenerating. David Riesman's *The Lonely Crowd,* William Whyte Jr.'s *The Organization Man,* and essays on the family by Talcott Parsons in the 1950s all warned of the dangers of "the growing homogenization of the sexes." Masculine and feminine sex roles were becoming increasingly similar, they argued. In the consumer, corporate world of the 1950s, middle-class men were losing their authority and becoming "feminized." More married middle-class women were joining the work force than before and thus were becoming more masculine. This perceived convergence of sex roles worried sociologists, who argued that "contrasting definitions of femininity and masculinity were the cornerstone of the family."[62] The M-F test linked abnormal sex-role behavior to the "pathological" condition of homosexuality in order to stigmatize both. Masculinity and femininity became important cultural markers used to shore up gender differences during a time in which these distinctions appeared to be receding. "The trend of much educational and popular thought during the last generation," Popenoe warned in 1930, "has been toward minifying the differences between the sexes."[63] The emergence of gay subcultures and single working women called into question the sanctity and centrality of the family in modern life. Eugenicists meant to ensure that marriage, procreation, and family would maintain their central position in American society.

Just as the Stanford-Binet test designated Terman as an expert on intelligence, the M-F test made him an expert on sex roles. His overall conclusion that "marked deviations from sex-appropriate behaviors and norms were psychologically unhealthy" underscored the eugenic argument that civilization would advance only on the feet of healthy children born of well-adjusted parents.[64] The further one deviated from the norm, the less likely he or she would succeed in marriage and parenthood.

This argument paved the way for Terman's next study, published only two years after *Sex and Personality.* In *Psychological Factors in Marital Happiness* (1938), Terman argued that the success or failure of a marriage depended largely on the personalities and the compatibility of husband and wife. Not surprisingly, Terman argued that marital stability was contingent on adherence to gender norms—in other words, masculine men and feminine women made the best partners. Applying the M-F test to the eight hundred married couples in his study, he found that happily married women were cooperative, did not "object to subordinate roles," and were "not annoyed by advice from others." They were "conservative and conventional" in religion, morals, and politics. Unhappily married women, however, were "egocentric," "overanxious," aggressive, and radical in politics and in morals.[65] His message was clear: traditional, conservative, feminine women would be rewarded with happiness in marriage; subversive women who challenged gender norms would not.

Terman's work attracted much positive attention.[66] One reviewer for the *Journal of Heredity* remarked that "all investigations of the problem of happiness and marriage will, in the future, be dated either 'before Terman' or 'after Terman.'"[67] The *Eugenics Review* noted that *Psychological Factors in Marital Happiness* was of "particular value to eugenists" and attributed its usefulness to Terman's ability to measure masculinity and femininity.[68] Dickinson was enthusiastic about Terman's work, telling him that "nothing whatever has happened to practical marriage studies more important than your book."[69] Although much of Ellis's emphasis on sexual compatibility in marriage was discredited in Terman's book, he too claimed it was a "valuable contribution" to the subject and recommended it to friends.[70]

Paul Popenoe: "The Man Who Saves Marriages"

Terman's interest in promoting traditional gender roles in marriage led him to the work of his friend and colleague Paul Popenoe. The two had been quite influential in eugenic-sterilization work; while Popenoe was the researcher for two major surveys of eugenic sterilization in California for Gosney's Human Betterment Foundation, Terman was a prominent HBF board member. Thus when Terman was looking for subjects for his marital-happiness survey, he turned to Popenoe for support. "At present in America about one marriage in five ends in divorce," Terman wrote to Popenoe in November 1934. "Will you not contribute your 'bit' to help us in our attempt to locate some of the causes of so much domestic unhappiness?"[71] Popenoe agreed to cooperate by selecting and administering Terman's marital-happiness survey to eight hundred Los Angeles–based couples affiliated with Popenoe's American Institute of Family Relations.

Though Popenoe had a Ph.D. in heredity and spent much of his early professional career breeding date palms, editing the *Journal of Heredity,* and researching the effects of eugenic sterilization in California for the HBF, he later claimed that his real passion was for the American family. "Nothing is more important to America today," he declared in a 1960 interview with the *Ladies' Home Journal,* "than the preservation of a sound family life."[72] Recognizing in the early 1930s that eugenicists might have more influence promoting propagation of the "fit" than restricting the "unfit," Popenoe convinced the wealthy Gosney to finance the American Institute of Family Relations.

Popenoe was right. While eugenic sterilization as a means of curtailing the "weaker" elements of the population peaked in the 1930s and then began a slow decline in the 1940s, positive eugenics, in the form of marriage and family counseling, proved more potent. Popenoe's Institute opened in February 1930 as "the first organized attempt in the United States to bring all the resources of science to bear on the promotion of successful family life."[73] By 1962, with a staff of seventy and seven branches in the Los Angeles area, the Institute claimed to have salvaged the marriages of some seventy-five thousand couples. The *Ladies' Home Journal* declared that the Institute was the "world's largest and best-known marriage counseling center" and boosted its fame by regularly featuring articles by Popenoe between 1942 and the 1960s.[74]

Marriage counseling, a new and innovative process, originated in Germany in the 1920s. In February 1926, the Prussian Ministry for Social Welfare called for the establishment of "officially medically directed marriage counseling centers to advise prospective mates and parents about their eugenic fitness for marriage and procreation."[75] By 1933, there were approximately three hundred centers nationwide, developed as alternatives to private birth-control clinics already in operation. When Marie Kopp (who would later survey the results of the German sterilization law and find them praiseworthy) conducted a survey of the German centers in 1932, she found that they stressed the importance of "marital fitness"; they believed that such fitness was essential in order for couples to "pass on a healthy mind and body to their children." Under Nazi social health policy, German marriage-counseling centers exclusively emphasized the importance of racial hygiene, encouraging larger "fit" families by offering baby bonuses and marriage loans, as well as targeting the "unfit" for sterilization.[76]

In the United States, marriage counseling was promoted by both eugenicists and birth-control advocates. Doctors Abraham Stone and Hannah Stone (both supporters of eugenic sterilization) opened a marriage-counseling service in 1931 after Hannah Stone had encountered numerous sex problems in her patients at Sanger's Birth Control Research Bureau. The couple offered premarital counseling (in part because of the support of eugenicists) and sex education, which included information about birth control. Planned Parenthood later published selections from their best-selling *A Marriage Manual* (1935) in a pamphlet on premarital counseling that emphasized the importance of sexual adjustment for a successful and stable marriage.[77]

Contributors to another popular marriage guide, *Successful Marriage: A Modern Guide to Love, Sex, and Family Life,* included a number of eugenic supporters, including Dickinson, Terman, and Popenoe. It was edited by Morris Fishbein, former editor of the *Journal of the American Medical Association,* who believed that the "mentally deficient must be prevented from reproducing" and that sterilization was the only solution. The authors reiterated Dickinson's emphasis on "preventive gynecology," arguing that marriage counseling was a form of preventive medicine in that it helped to create and sustain a "well-adjusted marriage and family."[78]

But it was in Los Angeles, where "the formula for modern marriage and family life radiated" from, that marriage counseling took on its most influential eugenic form, at Popenoe's Institute of Family Relations.[79] As director, Popenoe drew on the support of fellow eugenicists he had worked with during his years as a researcher for the HBF. Gosney not only provided financial assistance but also participated in Institute conferences. Roswell Johnson, sterilization advocate, former president of the AES, and coauthor with Popenoe of *Applied Eugenics,* became director of the department of personal service at the Institute. In addition, the director of the Eugenics Record Office sought permission in 1940 to forward any requests for marital advice to the Institute, a proposal that Popenoe gladly accepted.

More so than any of his colleagues, however, Popenoe achieved fame and popularity in the mainstream media. He became best known as "the man who saves marriages."[80] His *Ladies' Home Journal* series—which in the 1950s came to be called "Can This Marriage Be Saved?"—featured success stories of couples who had turned to the Institute for guidance. The *Journal,* which still publishes this column, calls it "the most popular, most enduring women's magazine feature in the world."[81] Popenoe also made his mark on radio and television; he was a favorite guest on "Art Linkletter's House Party," which debuted on radio in 1944 and on CBS television in 1952, and Popenoe even hosted his own television show. Popenoe invited couples who had filed for divorce in the city of Los Angeles to appear on his program and air their complaints. In this unscripted, unrehearsed drama, Popenoe sat in a judge's chair, listened to them argue, and then gave them "good sound advice, backed by the latest scientific research."[82] Though Popenoe's motive was still eugenic, its message and medium, popular marital advice channeled through women's magazines and television, effectively separated it from any negative, Nazi-based connotations. (My mother, who rarely picks up a women's magazine, still remembers Popenoe's smiling photograph and interesting stories about unhappy marriages miraculously transformed by marriage counseling in the 1960s.)

Though the medium may have masked the overt eugenic message, Popenoe did not hesitate to articulate his ideals in marriage counseling. In an address delivered at the Los Angeles Public Library in 1940, Popenoe explained, "Marriage is one of the fundamental institutions which has perpetuated the human race and made civilization possible."[83] While marriages had increased by 300 percent since the Civil War, divorces had increased by 2,000 percent. The implications for eugenics and for society were potentially devastating, he argued, for "survival of a civilization depends on its being 'family-minded.'" In order for a nation to survive, children must come from parents who are "mentally, physically, and emotionally sound"—and of course married.[84] "If, in the long run, the only wealth of a country is its children," he explained, "a moment's consideration reveals that much of what should be America's wealth is greatly depreciated."[85] Through education and counseling, mothers needed to practice reproductive morality. Like the fictional Janet March, they needed to recognize that their greatest purpose was to create the sound minds and bodies of tomorrow's children.

To spread the word that the American family was in trouble, suffering from a "deep-seated social maladjustment," Popenoe embarked on an educational campaign.[86] In the early 1930s, he exercised influence in three different capacities: as director of the Institute of Family Relations, chairman of the Southern California branch of the AES, and secretary of the HBF. He drew on the resources of all three organizations (the Southern California AES, for example, had important contacts with the Los Angeles public school system) to structure a series of lectures, conferences, pamphlets, and articles that would reach the greatest number of people in the Los Angeles area.

For example, in December 1933, the Institute sponsored an all-day regional conference on family relations at the Pasadena Presbyterian Church. The program included twelve different sessions on aspects of marriage, sexuality, eugenics, and the family and drew on the expertise of various professionals. Popenoe discussed "sex in everyday life." Gosney chaired a panel on heredity and eugenics, which included a discussion of both sterilization and the "means by which the families producing superior children can be encouraged"—an explicit acknowledgment that the Institute was interested in promoting marital happiness in order to encourage positive eugenics. Also included in the conference were two sessions on marriage and choosing a eugenically sound mate, a panel addressing the "need of a more normal social life" for young people, and a discussion of "sexual perversions and abnormalities."[87] Held in a church and promoting traditional family values and the sanctity of marriage, the conference assured community participants that applying scientific, eugenic standards to the institution of marriage would help to strengthen its place in society.

The Institute also sponsored conferences geared more specifically toward ensuring eugenically fit marriages. It regularly offered a premarital conference, whose purpose was to "eliminate those who, because of mental or emotional defect, are not qualified to marry successfully" and to promote the "intelligent selection and successful attraction of a mate" for those who were mentally and physically sound (figure 7).[88] The Institute encouraged every couple considering marriage to attend such a conference and included in it a rigorous process of testing to determine who should marry. The procedure included a study of the individual's personal history and of his or her family history, a physical examination, an interview, recommended reading, and instruction.

Central to the eugenic philosophy espoused by the Institute were the principles of masculinity and femininity articulated by Terman. As late as 1974, Popenoe used a flyer entitled "Some Biological Differences between Men and Women" in Institute workshops. Intended to shore up sex roles as distinct natural categories that ought not to be challenged, the pamphlet stressed that "men and women differ in every cell of their bodies. This difference in the chromosome combination is the basic cause of development into maleness or femaleness." While Terman's M-F test convinced one reviewer that "sex differences go deeper than the genitals," Popenoe located masculinity and femininity as deep as possible—on a cellular level. Women laughed, cried, and fainted more easily than men, while men were 50 percent stronger than women.[89] Popenoe suggested that to challenge these sex-specific attributes was to go against nature. "Equality does not exist in Nature," he reflected in 1930; "it is merely a concept of the mathematician. . . . Man and woman can never be equal." Foreshadowing the concerns of 1950s sociologists, Popenoe argued that

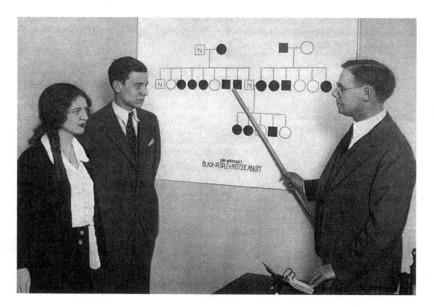

Figure 7. Paul Popenoe advising a couple wanting to get married on the importance of eugenics and family heritage. Courtesy of the American Philosophical Society.

the problems of the modern family stemmed from a "lack of needed emphasis on these [sex] differences."[90]

Popenoe was also deeply offended by homosexuality and, like Dickinson and Terman, viewed it as a pathological condition that threatened marital stability and eugenic purity. "Are Homosexuals Necessary?" he asked in the title of an Institute publication. Citing a study of 710 homosexual males who came to the New York Academy of Medicine for treatment and unanimously reported that they would "choose to be other than homosexual if it were their choice to make," Popenoe stressed that therapists should pressure homosexuals to change their sexual disposition. "It is an impossible basis on which to build family life," he stressed, and warned that the pathology persisted through the "continual recruiting of new victims" and thus threatened the stability of healthy heterosexuality.[91] The situation would best be remedied by responsible parenthood. Quoting the authors of a survey on homosexuality, Popenoe declared, "It is just about impossible for a homosexual to be the product of warmly loving, sensible parents and a sexually well-adjusted home atmosphere."[92] Reinforced by Terman's M-F study, this demonization of homosexuality and its association with inappropriate gender identity served to buttress the eugenic ideal of traditional gender roles and the importance of sound parenting.

Popenoe's plea to salvage the American family and to make American society in general more "family-minded"[93] found its largest audience not at the American Institute of Family Relations but in the *Ladies' Home Journal*. The magazine was an ideal setting for Popenoe's message; throughout the first half of the twentieth century, the *Journal* celebrated femininity and homemaking. In addition, as the top-selling women's magazine through the 1950s, it successfully courted white, middle-class, native-born women and became known as a "handbook for the middle class."[94] This was the very audience Popenoe and all eugenicists hoped to influence with positive eugenics. Eugenicists claimed that this group of "new women," who had chosen college and career before marriage and motherhood, were not reproducing enough to replace themselves.[95] If Popenoe could get the millions of *Journal* subscribers to think about eugenics and reproductive morality while planning their futures, his eugenic triumph would be complete.[96]

Popenoe began writing for the *Journal* in July 1941. Each article, based on case records from the American Institute of Family Relations, included a particular message about marriage for women. With titles such as "Make Your Quarrels Pay Dividends," "Give the First Baby Second Place," "Make Jealousy Work for You," and "Now Is the Time to Have Children," his articles prescribed subservient, affectionate, sexy, and domestic behavior for housewives—the very prescription given by Judge Lindsey's 1927 *Companionate Marriage*. In 1953, the *Journal* launched a "revealing new series on real-life marriages," entitled "Can This Marriage Be Saved?"[97] This series, based for decades on the counseling records of Popenoe's own Institute of Family Relations, continues to this day.

In his first article for the *Journal,* in 1941, Popenoe described the rocky marriage of the Reitzells, who found themselves increasingly incompatible during the Depression. Dismissing financial difficulties as the cause ("no marriage ever broke up because of financial difficulties—the real problem is in the personalities, not the pocketbooks," he claimed), Popenoe suspected that elements of weakness in the marriage needed to be addressed. Mrs. Reitzell had become too critical of her husband, who had lost his job and needed his home to be "an oasis in the desert, a haven of refuge," not a war zone. She was no longer feeling satisfied "merely to be Mr. Reitzell's wife and the mother of the Reitzell children; . . . she must be somebody on her own account." With counseling and the pursuit of hobbies to occupy some of her leisure time, Mrs. Reitzell began to come around. Like Janet March, she began to appreciate the importance of her role at home. "Being a homemaker is my first responsibility," she declared. "And to tell the truth, I don't particularly admire the woman who gives up her home for a job downtown, or the husband who lets his wife take a full-time job at the expense of her home and children." With help from Popenoe and the Institute of Family Relations, the Reitzells reestablished their home "on an emotional foundation which is apparently unshakable."[98]

Similar scenarios followed in later issues. After describing a particular couple's problem and its satisfactory resolution, Popenoe sometimes included a quiz or questionnaire that readers could complete to see how they would do in similar situations. "Are You Old Enough to Marry?" he asked after an article entitled "Marriage Is for Adults Only." Ten statements followed that required a "yes" to guarantee a woman success in marriage. They included: "You appear not to notice that your husband dented the fender of the car on his way home from the lodge meeting last night." "You are as polite to your husband as you would be to a perfect stranger." "You always say, and really believe, that your husband has helped you to grow up." Such adulation and politeness suggested more of a patriarchy than a partnership and helped to preserve a role for women based on subservience rather than self-esteem.[99]

Popenoe justified this role for women as one that would be rewarded with happiness in marriage. He pointed out that women who were college graduates had a divorce rate four times higher than that for men who were college graduates, a group that tended to marry their educational inferiors. The highly educated woman suffered the most because "she had 'independence' and 'self-expression' taught her so insistently, without any adequate guidance to attain it. Her attempts to deal unaided with this profound and obscure problem are too often based merely on infantile self-assertion. Of course they fail," he claimed. Dismissing women's attempts to attain equality in education, career, and marriage as selfish and childish, Popenoe argued that the "aggressive, overactive wife" was likely to be a discontented partner, while a wife "dominated by her husband is often quite happy."[100]

Popenoe promoted not only marriage but also motherhood in his *Journal* articles; his concern was not just that women remain in traditional gender roles but that they have more children to ensure that the future population would be dominated by the white, educated middle class. He found a high birthrate among this group particularly critical during periods of war, as did most eugenicists. Like disease, "feeblemindedness," and sexual immorality, war threatened the strength of the race by killing off its healthiest citizens. Like World War I, World War II generated anxiety about the effect of female promiscuity on the health of soldiers. The military again turned its attention to the "promiscuous girl." One journalist noted, "Fully 90 percent of the Army's [venereal] cases in this country are traceable to amateur girls—teenagers and older women—popularly known as 'khaki whackies,' 'victory girls,' and 'good-time Charlottes.'"[101]

The anxiety over "promiscuous girls" during World War II reflected more than a concern about venereal disease; many saw the emergence of these girls as indicative of a breakdown in the American family.[102] Lengthy separations of husbands and wives, a rise in the divorce rate, and the entrance of five million additional women into the labor force all strained families.[103] Wartime propaganda reminded soldiers that they were fighting for the family rather than out of a sense of political obligation. It also urged women to "turn their postwar energies to marriage and child rearing."[104] Eugenicists reinforced this propaganda: they believed it crucial that during wartime middle-class reproductive rates increase in order to offset the loss of some of the nation's strongest and healthiest men. Beginning in the postwar era, this profamily, pronatalist message had widespread popular appeal.

In "Now Is the Time to Have Children," Popenoe made a hard sell for immediate motherhood. Writing in wartime (1942), he reasoned that "the future uncertainties are one of the strongest reasons why people who can give children a good start in life should do so right now. The nation will need leadership and it must be furnished by young people coming on, unless the general desire is to leave an unoccupied continent for Japanese colonization as soon as the existing population can die conveniently." Regardless of war, he argued, the world will go on. "The question is," he asked, "whose children will guide it?"[105]

In case instilling a sense of reproductive morality in his readers was not enough to inspire motherhood, he appealed on personal grounds as well. Based on the analysis of thousands of case records from his Institute, he found a direct correlation between marital happiness and family size. The more children, the happier the (middle-class) parents.

He also appealed on a physiological level, arguing that "children are especially needed by the wife for her own health and mental hygiene. . . . A woman's body is made for childbearing and is not functioning normally unless it bears children." In a gush of pronatalism, Popenoe declared that "nothing is better than a pregnancy, with its powerful stimulus to all the glands of internal secretion, to complete the development of the reproductive system." Addressing a white, middle-class audience, he was able to make such a general claim for the emotional and physical benefits of reproduction even while advocating sterilization as a simple and healthy procedure that did not "unsex" for the unmarried and the poor. His claims about the physical and psychological benefits of motherhood was echoed in the reports of marriage and sex counselors in the 1950s and 1960s, including the well-known team of William Masters and Virginia Johnson.[106]

The cost to women of "doing too much" in high school could be great. Warning against the dangers of mental and physical overexertion for young women's bodies in their reproductive prime, he echoed the sentiments of G. Stanley Hall, who fought against female higher education at the turn of the century on the same grounds. Popenoe claimed that "many a girl nowadays carries too heavy a load in the high school period; she is using up her *vital capital* too rapidly in studies, work, social affairs, [and] athletics, just at the time when she needs all that biological reserve to bring about the thoroughgoing transformation of the reproductive organs which takes place during the years of adolescence and which is so much more extensive in a girl than in a boy." If eugenic, nationalistic, emotional, and physical justifications for early marriage and motherhood and for traditional gender roles were not enough, Popenoe offered biological reasons as well. His solution to the problem of an overwrought high school girl who had unknowingly interrupted the healthy development of her reproductive organs was "early marriage and parenthood." He urged his readers to contribute to the birthrate and join those mothers who "believe in themselves, believe in America, and believe in babies."[107]

Perhaps the article that was most effective in encouraging reproductive morality in *Journal* readers was a 1943 photo essay entitled "Meet an Engaged Couple." Dot, a college student, and Dick, a member of the Air Force, agreed to have their relationship analyzed in the *Journal* by Popenoe, who flew to Ohio to determine their chances of marital success. A full-page close-up photograph of the young couple embracing illustrated the *Journal*'s conception of the ideal couple. Dick is leaning over Dot in a tight embrace, looking down at her firmly but lovingly, while a "U.S." military pin gleams brightly under his chin. Dot gazes admiringly into her fiancé's eyes, wearing feminine clothes and a moderate amount of makeup to appear attractive yet conservative.[108]

The following pages visually trace Dot and Dick taking part in the premarital counseling that Popenoe offered regularly at the Institute of Family Relations. Popenoe immediately puts the young couple at ease with a warm smile and a handshake. Then Dick and Dot are each assigned a counselor who will determine their emotional maturity and their readiness for marriage. Again, counselors and couple appear relaxed and friendly. Next, a physician examines Dick for health complications that could affect marriage and children. Photos of both candidates with their families follow, illustrating that they both come from stable and reputable backgrounds and get along well with their parents.[109]

Photos of Dot underscore her femininity. She is pictured in a college cooking class practicing her housekeeping skills. "Those early-marriage blues would seldom get brides down if they came to marriage with some housekeeping skills," Popenoe writes. "What other job would you walk into without training, and how long would you keep it?" Dot had already begun to practice cooking and remarked that "nobody seemed to die" from it. Popenoe adds, "Girls who go to college often try to assert their individuality in marriage. Result: their divorce rate is four times higher than that of college men." Yet Dot appears safe from such dangerous individuality. Stephens College girls "make good wives," as only 3 percent of their alumnae get divorces (compared with 13 percent of all female college graduates). The college had a reputation for "training future wives and mothers to be efficient and womanly women"—a eugenic ideal.[110]

Next, we see Dot surrounded by young schoolchildren, an image designed to evoke her fertility and her role as a mother. As a child-study major who worked in a kindergarten classroom during college, she appears a ready and willing candidate for motherhood. The final photo shows her college home and family professor, Dr. Henry Bowman, pointing to a chart of the female reproductive system and fetal development. Bowman, author of *Marriage for Moderns,* offered the "most successful college course in the United States," which was based on material acquired while he was a guest counselor at Popenoe's Institute of Family Relations.[111] Dot appears to be listening earnestly to his words of wisdom. Underneath, Popenoe writes that this professor instructs many young women on problems of love and sex and asks them of their potential mates, "Do you love him faults and all?" Men must be accepted as they are, while women must be willing to accommodate; "trying to reform a man is an infantile reaction," Popenoe explains, "serving only to bolster the girl's ego."[112]

Popenoe declared a healthy prognosis for Dick and Dot's marriage. Juxtaposing the feminine, relaxed, and radiant images of young Dot with warnings about what happened to women who did not adhere to the domestic mold, he presented a powerful message to *Ladies' Home Journal* readers. True happiness and fulfillment would reward women like Dot, who did not challenge her role as wife and mother but accepted it willingly. The message was hardly new but was disguised as a scientific approach to marital happiness; Popenoe gained millions of followers and the *Journal*'s acknowledgment that "marriage never had a more ardent champion."[113]

Postwar studies reinforced Popenoe's argument that women needed to focus on procreation rather than education. *Newsweek* commented in 1946 that "for the American girl books and babies don't mix," claiming that scientists had recently discovered that the "higher-educated wife . . . brings down the birth rate."[114] The *American Journal of Sociology* published that same year a study of married women in the labor force, which also linked women's education and employment to the declining birthrate. But times were beginning to change. By 1947, the birthrate was up to 25 percent (from a low of 18.7 percent in 1935) and was stabilizing. Motherhood was becoming "an absorbing, creative profession—a career second to none," according to one journalist in 1946.[115]

And, indeed, the 1950s witnessed an unprecedented celebration of motherhood and domesticity. As one historian notes, "The 1950s elevated family life to a level of sacredness never before witnessed in our history." A new notion of "family togetherness" emerged along with the proliferation of family goods: the family car, the family restaurant, and the family film. Much of the emphasis on family was a result of the baby boom of 1946–64, when 76.4 million babies were born in the United States, with more than four million per year arriving between 1954 and 1964. Significantly, and to the delight of eugenicists, the highest jump in fertility occurred among well-educated white women with "medium to high incomes."[116]

Historians acknowledge the significance of the baby boom, "perhaps the most amazing social trend of the postwar era," and its impact on American society and culture. Yet while scholars offer a range of explanations, there is no consensus on the causes of the boom. Some point to wartime propaganda by government agencies aimed at raising the birthrate and returning women to the home. Another hypothesis is that Americans were on a quest for "normalcy" after the disasters of war. Neither of these can explain why the birth and marriage rates began to rise before the war ended, nor why the boom lasted for so long; after all, previous wars had not caused such a demographic wonder. David Patterson, dismissing other theories, argues that the younger parents of baby boomers "seemed to yearn for marriage and children," perhaps as a way of coping with the "pressures of an increasingly complex and bureaucratic world."[117]

But where did such yearning originate? Why were marriage and motherhood viewed as an effective solution to the problems facing modern America? Reasons are undoubtedly complex; clearly not all women who gave birth during the baby-boom years were simply duped by fear or propaganda. Yet the cultural climate of the 1930s and 1940s, particularly the emphasis on marriage and parenthood promoted by the positive-eugenics campaign, clearly played a role in the procreative choices women of the baby-boom years made. "To be a mother was very much more vivid, important, thought-about, and valued at the height of the baby boom than to be a wife," John Modell explains.[118]

Motherhood outside of marriage, however, was strictly forbidden. Rickie Solinger's study of unwed motherhood during the postwar era testifies to the cultural concern that unrestrained female sexuality, symbolized by premarital pregnancy, jeopardized the American family. "The label of illness—along with the virtual expulsion from the community—experienced by white unmarried mothers warned like women everywhere to keep their commitment to marriage and motherhood strong, in that order, and reminded them—indirectly, but powerfully—of the wages of unsanctioned sexuality," Solinger notes.[119]

Even married mothers posed a potential threat if they did not adjust "properly" to their roles. Maladjusted wives and mothers, in Solinger's words, "undermined the possibility of the harmonious nuclear family."[120] As demonstrated in the works of Dickinson, Terman, and Popenoe, marital adjustment was of enormous concern to those invested in preserving and stabilizing the family.

In the 1950s, the widely popular *Ladies' Home Journal* in collaboration with Popenoe continued to play a formative role in spreading the word about the importance of marital adjustment and parenthood. According to the editors of the *Journal,* when the column entitled "Can This Marriage Be Saved?" debuted in January 1953, "it immediately captured the attention of millions of readers." Sonya Friedman notes that the column "not only is the most popular [as well as the longest running] women's magazine column of all time, but it has also served to legitimize marital counseling in this country." Each month, readers learned of a couple experiencing marital difficulty who had sought the help of Popenoe's American Institute of Family Relations. Frequently, the couple had already filed for divorce but sought marriage counseling from the Institute at the suggestion of a lawyer or judge. "Can This Marriage Be Saved?" consisted of three parts: the wife's view of the problem, followed by the husband's view, and, finally, the analysis offered by a marriage counselor from the Institute. At the top of the front page of the column was Popenoe's picture, along with an introductory paragraph written by Popenoe explaining the perils of marriage in modern America. With marriage counseling, however, almost any marriage could be saved; Popenoe boasted of an 80 percent success rate at his institute.[121]

Issues leading to marital tension in the 1950s were similar to the ones analyzed by Popenoe in the 1940s. What was effective about the new column was its use of "real people in real-life dramas." Its popularity, editors claim, stems from the fact that "millions of readers see themselves and their marriages reflected in its pages." Female readers identified with wives who struggled with their marriages. In the 1950s, wives were not fighting for independence or autonomy (as they sometimes were in the examples in the 1940s) but frequently for husbands to recognize and appreciate their role as mothers. "I wanted a baby practically from the first, but Dan wasn't interested in acquiring a family," complained Lucy in April 1954. "I want my babies while I'm young, but my husband seems determined to postpone having a family until he's an old man," Diana revealed in the premier 1953 column. In both these marriages, tension subsided after the couple worked with an Institute counselor and after the arrival of children.[122]

The May 1954 column dealt with a topic of keen interest to family promoters: the problem of infertility. "Several million marriages in the U.S. are childless, contrary to wishes of husband and wife," wrote Popenoe. In many cases, he continued, the cause was not physiological but was some form of "emotional disturbance." With the help of marriage counseling from the Institute, however, many couples discovered that successful marital adjustment resulted in fertility. Kate and Joe, featured in this article, were one such couple. They had been trying unsuccessfully for over three years to have a child. Kate had undergone several tests (there is no mention of Joe's ever having been examined), and she was finally told by doctors that the problem was psychological. Kate was "tense and nervous" around her husband, who, instead of enjoying evenings at home with his wife, "thinks of our home only as a way station and a place to play the fool. He changes his clothes there so we can rush somewhere else," even though she spent her days "cleaning the house from top to bottom." In addition, Kate suffered from sexual maladjustment, admitting that "for Joe and me, sex is an ordeal. It always has been. Sex causes me pain, extreme pain. Joe is not to blame. It's all my fault. According to the doctors, my nervous system is responsible for the trouble." After working with the Institute counselor to overcome her fear of sexuality, Kate became pregnant.[123]

According to Margaret Marsh and Wanda Rommer, the notion that infertility was frequently a result of psychological rather than physiological disorders became a "newly fashionable idea" in the 1950s, when some doctors estimated it was the cause in up to 75 percent of all cases. This new theory of infertility appeared not only in medical journals, such as *Fertility and Sterility,* but in popular magazines, such as the *Ladies' Home Journal.* Infertile women were sometimes blamed for their sexual maladjustment, their "lack of femininity," an "unconscious rejection of motherhood," or refusing to give up their careers. One psychiatrist told his colleague about an infertile couple who discovered through psychoanalysis that they had conflicts "dealing with repudiation of masculinity and femininity." Soon after this problem was recognized and dealt with, the wife became pregnant. Fear of the "growing homogenization of the sexes," expressed initially by Popenoe and Terman, was affirmed in the infertility studies of the 1950s. Without healthy adjustment to gender-appropriate behavior within marriage, couples allegedly risked infertility. Marital maladjustment, like syphilis, was a disease that without proper treatment could result in sterility and thus could truly debilitate the race.[124]

Marital maladjustment notwithstanding, pronatalism persisted throughout the 1950s. Though the baby boom peaked in 1957, the "cultural imperative for large, planned families" did not lessen until the mid-1960s, when the birthrate began to drop.[125] When asked to state the ideal family size, a majority of women surveyed before the war replied two children. In the 1950s, the most common response to this question was four children. By 1961, the ideal family size for many was up to five children. A *Ladies' Home Journal* poll of readers revealed that most women wanted four children and "many" wanted five.[126] The women polled in the *Journal* undoubtedly also paid close attention to "Can This Marriage Be Saved?" While pronatalism was a common facet of society in the 1950s, it had emerged in the positive-eugenics campaign of the 1930s and 1940s. "Marriage is not complete without children," one housewife commented in 1955.[127] Popenoe could breathe a sigh of relief. His message to white middle-class women had been heard.

The tactical shift of eugenicists in the 1930s that allowed for an environmental approach also permitted positive eugenics to take hold as a tenacious and popular ideology in postwar America. Eugenicists such as Dickinson, Terman, and Popenoe moved beyond the prevention of the "unfit" to the promotion of white middle-class procreation. Providing new scientific evidence that nineteenth-century gender roles were not useless relics of Victorianism but essential elements of modern society, they offered a new rationale for separate spheres. As Terman quantified masculinity and femininity and Popenoe applied it to marriage counseling, new standards of social and sexual normality emerged. Stigmatizing those who deviated from the "norm" as pathological, eugenicists reinforced masculinity and femininity as healthy, well-adjusted roles for modern Americans. Their vision of a modern morality, which I call reproductive morality, was little more than a nineteenth-century notion of gender dressed in the garb of science.

But it worked. Popenoe's plea for society to become "family-minded" was answered in the pronatalist culture of the 1950s.[128] The "passion for parenthood" that is most commonly linked to Cold War culture had its roots in the positive-eugenics campaign that began in the 1930s and grew in influence during the 1940s and 1950s. While postwar abundance supplied the means for achieving suburban bliss in the 1950s, the impetus behind the move to suburbia and the new emphasis on baby-boom family culture stemmed in part from the influential promotion of procreation by eugenicists.[129]

In their search for effective eugenic solutions to the decline of Victorian morality in early twentieth-century America, eugenicists tried a number of strategies. Initially focused on segregating "moron girls," who challenged nineteenth-century notions of virtue, they switched to sterilization as a more effective and affordable solution when they realized how widespread the problem of female sexual "delinquency" had become. Finally, armed with a new justification for eugenics as a cure for cultural as well as biological ills, they targeted mainstream America in their positive-eugenics campaign to build a better race. The "golden age" of eugenics occurred long after most historians claim the movement had vanished. The baby boom of the 1950s represented the triumph eugenicists had been looking for. In their campaign, eugenicists such as Terman and Popenoe ensured that Janet March and not Ann Cooper Hewitt or Stella Dallas would represent womanhood in twentieth-century America. Faced with alternatives to marriage and motherhood, the fictional Janet March came to believe, as did many real women between the late 1920s and 1960, that her true contribution to society lay in her potential to procreate.

Epilogue
Building a Better Family

In the 1989 movie *Parenthood,* directed by Ron Howard, Keanu Reeves plays Tod, the teenage son-in-law of middle-class divorcee Helen (played by Dianne Wiest). About to be a father himself, Tod comments on the negligence of Helen's first husband, who has abandoned Helen's children to raise a new family with a younger wife. "You know, Mrs. Buckman," Tod states thoughtfully, "you need a license to buy a dog or drive a car—hell, you even need a license to catch a fish. But they'll let any butt-reaming asshole be a father."[1]

The irony in Tod's statement, of course, is that he is hardly a role model for parenthood; an unemployed, high school–dropout, race-car driver living in his mother-in-law's house, he is portrayed as unlikely to do much better than Helen's ex-husband did. Yet movies and television series of the late 1980s and 1990s suggested that the breakdown of the modern nuclear family is to be laughed at as much as mourned. Sitcoms on network television from this period, including *My Two Dads, Will and Grace, Grace under Fire, Friends, Ellen,* and *Murphy Brown,* critiqued the traditional family structure by offering alternatives that elicit laughter but also exhibit some success. Here, children are raised with anywhere from two fathers to no parents; major characters such as Ellen have come out of the closet; and others negotiate parenting with lesbian ex-wives and their new partners.

Such programs appear to affirm the claim of many historians and sociologists that the traditional family system of the 1950s broke down as early as the 1960s. "Scholars agree that family and reproductive behavior in the postwar period were aberrant, a 'last-gasp orgy of modern nuclear family domesticity,'" Wini Breines remarks. "Demographic patterns in the 1950s do not conform to long-term historical trends; they constitute deviant blips on the charts."[2] Scholars cite the sexual revolution, the women's liberation movement, gay rights, reproductive rights, the rise of individualism, and economic changes as, for better or for worse, effectively undermining traditional notions of marriage and family and leading to a rise in divorce rates and a decline in marriages and births.

However, Tod's comment on fatherhood and its resonance in contemporary America suggest that the revolutionary nature of the 1960s was not enough to wipe out the powerful cultural messages about marriage and family promoted throughout the first half of the twentieth century. The teenage boy's observation that Americans need a license to do anything except parent both echoes the concerns of eugenicists and foreshadows the arguments of current "family-values" campaigners. Robert Dickinson remarked in 1928 that "examination for fitness has become customary for all occupations save marriage and parenthood. Eventually common sense may be expected to demand a similar preparation before deciding on matters so important to the life of the individual and the race." Sociologist David Popenoe (son of Paul Popenoe) is among numerous family-values campaigners currently advocating a number of measures to ensure that children are born only to well-adjusted married parents; these measures include "getting people to select mates wisely and to marry and have children only when they feel ready."[3]

In fact, the subject of *Parenthood,* the complications of the "postmodern" family, became a major theme in both popular culture and politics in the late 1980s. "Reminiscent of the early baby-boom days," Elaine May notes, "babies, children, and parenthood began to permeate the nation's popular culture." A "new pronatalism" was ushered in by a media blitz targeting middle-class career women as in danger of becoming infertile. According to a study published in the *New England Journal of Medicine* in 1982, women's ability to conceive dropped significantly after age thirty. Policymakers and the press frequently targeted the women's movement and abortion rights as bearing responsibility for this "epidemic" of infertility: too many women were waiting too long before trying to start a family. In *Backlash: The Undeclared War against American Women,* Susan Faludi argues that the "infertility epidemic" was not a medical problem but a "political program."[4]

Some analysts described the situation in eugenic terms. In 1987, Ben Wattenberg, senior fellow at the American Enterprise Institute, published

The Birth Dearth, which he then promoted in his newspaper column, speeches, television appearances, and radio talks. "I believe the demographic and immigration patterns inherent in the Birth Dearth will yield an ever smaller proportion of Americans of white European 'stock'... and this will likely cause more ethnic and racial tension and turmoil than would otherwise occur," he warned. He, too, accused the women's movement of encouraging women to postpone marriage and motherhood, to seek professional advancement, and to insist on reproductive rights. Richard Herrnstein, who in 1994 would coauthor the controversial (and blatantly eugenic) bestseller linking race with intelligence, *The Bell Curve,* also blamed "brighter" women for choosing education and careers over children.[5]

Wattenberg's "birth-dearth" slogan was immediately adopted by the New Right in their "family-values" campaign. They unleashed the threat of "cultural" or "genetic" suicide, claiming that women's liberation had caused the birthrate decline—a major emphasis in the political platforms of Pat Robertson, Pat Buchanan, and Dan Quayle. At the same time, they warned of a second, related epidemic: the alleged rise of illegitimate black births. Both these assertions were false: the black illegitimacy rate was falling, while the national fertility rate did not fall in the 1980s but held constant at 1.8 children per woman (where it had been since 1976). In fact, the population was increasing by over two million people a year— more than in any other industrialized country.[6]

Regardless of the weakness of their claims, the New Right succeeded in pushing the Republican party further to the right, ensuring that it developed a platform that was antifeminist, antigay, and antiabortion. Putting "family values" at the center of its campaign, the New Right identified itself as "pro-life, pro-chastity, and pro-motherhood." Their presidential candidate, Ronald Reagan, opposed the Equal Rights Amendment (the first president to do so since it had been passed by Congress) and supported a "Human Life Amendment" that banned abortion and some types of birth control.[7]

But by the time of the 1992 Republican national convention, it appeared that the radical right wing had simply gone too far in their family-values campaign. Extremists such as Robertson accused feminism of encouraging women to leave their husbands and kill their children. When Quayle criticized sitcom heroine Murphy Brown (played by Candice Bergen) for glamorizing unwed motherhood, she struck back by denouncing Quayle in her series as "being out of touch with the problems of *real* families." A majority of voters apparently agreed, siding with Bill

Clinton's claim that the economy, not the family, was most in need of immediate reform.[8]

For a brief moment, it appeared as if Republicans had driven family-values rhetoric into the ground. But in April 1993, just a few months after Clinton took office, the liberal *Atlantic* announced on its cover that "Dan Quayle Was Right." With this controversial cover story, author Barbara Dafoe Whitehead, codirector of the newly established think tank the Institute for American Values, helped to trigger what sociologist Judith Stacey calls the "neo-family-values campaign." While continuing to blame America's social and economic problems on family decline, this revisionist campaign immediately distanced itself from the New Right, which peaked in the 1980s and then suffered a decline in 1992. Supported primarily by academics, this "bipartisan consensus" emphasizes "postfeminism" rather than "antifeminism" and "grounds its claims in secular social science instead of religious authority."[9]

As Stacey argues in her book, *In the Name of the Family: Rethinking Family Values in the Postmodern Age,* the neo-family-values campaign is far more influential than the "old-fashioned family values warriors" of the 1980s. In the late 1980s and early 1990s, "an interlocking network of scholarly and policy institutes, think tanks, and commissions" developed a new, national consensus on family values that "rapidly shaped the family ideology and politics of the Clinton administration and his New Democratic party." Central to this campaign are the Institute for American Values, directed by David Blankenhorn and Whitehead, and its affiliated research organization, the Council on Families in America, cochaired by Jean Elshtain and David Popenoe.[10]

The Institute for American Values maintains that it "ignored the labels of 'left' and 'right,' seeking instead to bring together good people from across the political spectrum and across the human sciences." This think tank claims to have "influenced and at times framed the national debate on fatherhood, marriage, family, and civil society in the 1990s." By 1996, the Institute optimistically declared that the "family-values" debate was resolved. "Almost everyone now seems to agree. Quayle was right."[11]

But sociologists Judith Stacey and Stephanie Coontz are among a growing number of academics speaking out against the "false consensus" promoted by such think tanks. Profamily academic centrists claim to offer objective evidence that the decline of heterosexual marriage and an "explosion" in the rate of unwed motherhood suggest, in the words of David Popenoe, that American society "could be on the verge of committing social suicide." Stacey and Coontz counter such claims by revealing how they have been oversimplified by centrists, as well as by exposing some of their research flaws and challenging many of their assumptions.[12]

Their battle, however, is not an easy one. "New-consensus" organizations have flooded politicians and the media with profamily reports and position papers. Many social scientists, feeling displaced by a more liberal feminist voice in academia, "have found the centrist campaign a route to considerable public influence, media celebrity, and even academic attention," according to Stacey.[13] Taking center stage in this media blitz is David Popenoe.

Popenoe is the fourth and youngest son of eugenicist, sterilization advocate, and alleged marriage saver Paul Popenoe. Former chair of the board of his father's American Institute of Family Relations, he is currently a professor of sociology and associate dean of social and behavioral sciences at Rutgers University, founder and cochair of the Council on Families in America, and author of several books on the modern family, including *Life without Father: Compelling New Evidence That Fatherhood and Marriage Are Indispensable for the Good of Children and Society* (1996).

Popenoe describes himself as a middle-of-the-road New Democrat and claims that most of the twenty members of the Council on Families in America fit this description (though Stacey notes that more of them are conservative than liberal). As a prominent leader of the neo-family-values campaign, he has voiced his concerns about the decline of the American family in numerous prominent settings, including the *Mac-Neil-Lehrer News Hour,* the *New York Times,* the *Chronicle of Higher Education,* the U.S. Department of Transportation, and numerous academic conferences. At the 1994 American Sociological Association annual conference, Popenoe attacked a fellow sociologist's interpretation of the current family situation, dramatically taking over the podium to decry the "liberal intelligentsia" as out of touch with popular interests.[14]

Popenoe tirelessly campaigns for the "reestablishment of fatherhood and marriage as strong social and cultural realities." Though he is careful not to attack feminism, he blames much of the decline of the modern family structure on the "broader cultural shift toward a radical form of individualism" that took off in the 1960s (and that included the "increased education of women"). The results, he argues, have been devastating; America has fostered a "culture of divorce" as well as a "culture of nonmarriage." The absence of fathers has led to increased juvenile delinquency and crime (60 percent of rapists come from fatherless families, he notes), "promiscuous girls," early childbearing, child abuse, and violence against women.[15]

Fatherless families are not only bad for women and children, he argues; they are bad for men as well (a message adopted by the Christian men's Promise Keepers movement and Louis Farrakhan's 1995 African American Million Man March). "Men who do not father and who are not married," Popenoe asserts, "can be a danger to themselves and to society." While motherhood comes "naturally" to women, men have to adapt to fatherhood, which acts as a "civilizing force." Drawing on the work of evolutionary genetics and psychology, he argues that males have the instinctive incentive to "spread their abundant sperm more widely among many females . . . to further solidify [their] chances of genetic posterity." This is particularly true if a male lacks "paternity confidence," or the assurance that offspring are his, for there is "no genetic advantage to him for investing in another man's child." Therefore, it is crucial to contain sexuality and procreation within monogamous marriage.[16]

Because Popenoe's message seems relatively innocuous (most people agree that values are important and that two parents are in most situations better than one), it is quite powerful. He is careful not to point a finger at feminism or at women directly and backs his argument with examples of problems that typically resonate with many women: domestic violence, sexual abuse, and economic disparity in today's society. Compared with the blatant antifeminist, antichoice, and anti-gay-rights platform of the New Right, Popenoe's agenda appears more liberal, more scholarly, and more objective.

As his father's son, however, David Popenoe and his considerable public influence remind us that many eugenicists of the early and mid-twentieth century wielded a similar sort of power and authority. Citing statistics that the American family was in decline, eugenicists promoted eugenic sterilization, marital stability, and a return to traditional gender roles as strategies for building a better race. Foreshadowing the attacks of neo-family-values campaigners, they also blamed a culture of individualism and self-gratification (this time in the 1920s) for weakening family bonds and focused on the Cooper Hewitt sterilization case as evidence. As academics and social scientists, they too commanded the attention and respect of popular and professional audiences.

In fact, much of David Popenoe's argument is merely a continuation of his father's crusade. In 1930, Paul Popenoe addressed the perceived problem of the low birthrate and marriage rate of female college graduates in the *Journal of Home Economics*. He proposed restructuring the college curriculum in order to ensure that every course addressed the importance of marriage and parenthood since "students lack an understanding of the place of the family in society," an understanding important both "socially and racially, to provide a citizenship that will work and vote intelligently for the conservation of the family."[17]

David Popenoe believes that marriage preparation "should be included in the curriculum of every high school in the land. Is there anything more important that could be taught to students this age?" he asks.[18] In 1995, the Council on Families in America, cochaired by Popenoe, released *Marriage in America: A Report to the Nation,* with the goal of promoting marriage and family. "Due to the extraordinary national attention generated by the report," the Institute for American Values claims, more than fifty thousand copies were distributed or sold. Among the proposals made to achieve its goals, the Council suggested to teachers, principals, and education leaders that they "eliminate the implicit and frequently explicit antimarriage bias currently prevalent in many school curricula." Just as Paul Popenoe had suggested sixty-five years earlier, the Council argued that teachers should "promote education for successful marriage as a regular part of school curricula."[19]

In 1997, the Council on Families released an additional report on college-level marriage and family-life textbooks, a report that received widespread media coverage, including articles and reviews in the *Los Angeles Times,* the *Chicago Tribune, U.S. News and World Report,* and the *Chronicle of Higher Education.* The Council condemned current texts as a "national embarrassment," in that they furthered the family's decline rather than strengthening it. According to the Institute for American Values, this report has "already encouraged textbook writers and publishers to improve the content and quality of these books." In addition, the Institute is putting together an anthology of readings on marriage that can be used in college courses.[20]

Both father and son also argue that a central component in the marriage and family campaign is the effort to maximize, rather than minimize, gender differences. "Man and woman can never be equal because they are different in every cell of their bodies," argued Paul Popenoe in 1930. "I believe lack of needed emphasis on these differences is one of the most important underlying causes of sex antagonism, failure to marry at all, or failure to make a success of marriage." His son agrees. "Despite their many similarities," he writes, "males and females are different to the core. They think differently and act differently."[21]

David Popenoe is careful, however, in how he presents his notion of gender differences. He does not believe that the solution is to return to the traditional roles of the modern nuclear family promoted in the 1950s.

Instead, women should temporarily return home, taking a leave of absence from their careers, in order to be the primary caretakers of young children. "But we do need to be careful," he stresses, "not to drift too far in the direction of parental androgyny," a lack of adherence to traditional masculine and feminine roles. Fathers and mothers are biologically different, he reasons, and thus they contribute different parenting styles to children, both of which are necessary. Mothers are more nurturing, while fathers set limits and encourage physical activity. "We should disavow the notion that 'mommies can make good daddies,' just as we should disavow the popular notion of radical feminists that 'daddies can make good mommies,'" he asserts. Parental androgyny is unhealthy not only for children but also for the parents. Just as Lewis Terman had argued in his marital-happiness study of 1938, David Popenoe currently claims that parental androgyny leads to marital breakup.[22]

The influence of the neo-family-values campaign, even more so than the rise of the New Right in the 1980s, suggests that the concerns and ideals that were promoted by eugenicists during the first half of the twentieth century and that culminated in 1950s culture did not disappear with 1960s liberalism. David Popenoe has helped to carry forward his father's campaign, situating it within a modern, "postfeminist" context. From father to son, the dream of strengthening the American family—by eliminating alternatives to the nuclear-family system that realistically address the serious problems of contemporary society—lives on. "In order to restore marriage and reinstate fathers into the lives of their children," writes David Popenoe, "we are somehow going to have to undo the cultural shift toward radical individualism and get people thinking again in terms of social purposes." Paul Popenoe might have added the term *racial* to his son's "social purposes." David warns that our society "could be on the verge of committing social suicide"; his father would have termed it "race suicide."[23] But the message is essentially the same. What America needs is a new, modern form of morality—a reproductive morality—to counter the ever-increasing individualism in society and to build a better family.

Notes

Introduction

1. Mabel Potter Daggett, "Women: Building a Better Race," *World's Work* 25 (1912–13): 228.

2. John D'Emilio and Estelle Freedman, *Intimate Matters: A History of Sexuality in America* (New York: Harper & Row, 1988), 172; Beatrice Hinkle, "Women and the New Morality," in *Our Changing Morality: A Symposium,* ed. Freda Kirchway (New York: Boni, 1924), 249; V. F. Calverton (1928), quoted in Nancy Woloch, *Women and the American Experience* (New York: McGraw-Hill, 1992), 397; "Sex O'clock in America," *Current Opinion,* August 1913, 113–14, quoted in James McGovern, "The American Woman's Pre–World War I Freedom in Manners and Morals," *Journal of American History* 55 (1968): 316.

3. Isabel Leavenworth, "Virtue for Women," in *Our Changing Morality: A Symposium,* ed. Freda Kirchway (New York: Boni, 1924), 102.

4. Margaret Sanger, "Why Not Birth Control in America?" *Birth Control Review,* May 1919, 10–11, quoted in Linda Gordon, *Woman's Body, Woman's Right: Birth Control in America* (New York: Penguin Books, 1990), 277.

5. Charles Goethe to Lewis Terman, 19 September 1947, Lewis Terman Papers, SC 38, 4–12, Stanford University Library, Department of Special Collections, Stanford, Calif.

6. Between 1907 and 1920, the annual average was 230 sterilizations. Edward J. Larson, *Sex, Race, and Science: Eugenics in the Deep South* (Baltimore: Johns Hopkins University Press, 1995), 28.

7. Frank Dikötter, "Race Culture: Recent Perspectives on the History of Eugenics," *American Historical Review* 103, no. 2 (April 1998); Gunnar Broberg and Nils Roll-Hansen, eds., *Eugenics and the Welfare State: Sterilization Policy in Denmark, Sweden, Norway, and Finland* (East Lansing: Michigan State University Press, 1996); Mark B. Adams, ed., *The Wellborn Science: Eugenics in Germany,*

France, Brazil, and Russia (Oxford: Oxford University Press, 1990); Nancy Leys Stepan, *"The Hour of Eugenics": Race, Gender, and Nation in Latin America* (Ithaca, N.Y.: Cornell University Press, 1996); Larson, *Sex, Race, and Science;* Martin Pernick, *The Black Stork: Eugenics and the Death of "Defective" Babies in American Medicine and Motion Pictures since 1915* (New York: Oxford University Press, 1996); Ian Robert Dowbiggin, *Keeping America Sane: Psychiatry and Eugenics in the United States and Canada, 1880–1940* (Ithaca, N.Y.: Cornell University Press, 1997), 238.

8. Stepan's *"The Hour of Eugenics"* is one of the few studies of eugenics that takes gender into account.

9. Larson, *Sex, Race, and Science,* 2; Stepan, *"The Hour of Eugenics,"* 108.

10. Hazel Carby, *Reconstructing Womanhood: The Emergence of the Afro-American Woman Novelist* (New York: Oxford University Press, 1987), 18; see also Stepan, *"The Hour of Eugenics,"* 135–36.

11. For historians who argue that eugenics was in serious decline by the 1930s, see Mark Haller, *Eugenics: Hereditarian Attitudes in American Thought* (New Brunswick, N.J.: Rutgers University Press, 1963), 179; Carl Degler, *In Search of Human Nature: The Decline and Revival of Darwinism in American Social Thought* (New York: Oxford University Press, 1991), 150–53; Gordon, *Woman's Body, Woman's Right,* 272. In *Sex, Race, and Science,* Larson agrees that "mainline" eugenics had lost credibility by 1935, although the rank-and-file eugenics of the South lingered, largely because of a "lag," which, Larson claims, "slowed the departure of eugenics from the region" (119–20). Daniel J. Kevles's *In the Name of Eugenics: Genetics and the Uses of Human Heredity* (Berkeley: University of California Press, 1985), a valuable study of eugenics in Britain and the United States, argues that by the 1930s mainline eugenics "had generally been recognized as a farrago of flawed science" (164). In its place, Kevles argues, emerged a "reform eugenics," cleansed of its earlier extremism and racism by changes in both ideology and membership. For those who challenge the assumption that eugenics either was in disrepute or was completely reconstructed, see Stepan, *"The Hour of Eugenics,"* 195; Stefan Kühl, *The Nazi Connection: Eugenics, American Racism, and German National Socialism* (Oxford: Oxford University Press, 1994); Barry Alan Mehler, "A History of the American Eugenics Society, 1921–1940" (Ph.D. diss., University of Illinois at Urbana-Champaign, 1988). For those who warn against equating eugenics with Nazi Germany, see Stepan, *"The Hour of Eugenics,"* 4–5, and Dowbiggin, *Keeping America Sane,* 239; on German eugenicists during the Weimar Republic who refuted the work of the Nazis, see Atina Grossman, *Reforming Sex: The German Movement for Birth Control and Abortion Reform, 1920–1950* (New York: Oxford University Press, 1995), vii.

Chapter One

1. Ben Macomber, *The Jewel City* (San Francisco, 1915), 12, quoted in George Starr, "Truth Unveiled: The Panama Pacific International Exposition and Its Interpreters," in *The Anthropology of World's Fairs: San Francisco's Panama Pacific International Exposition of 1915,* ed. Burton Benedict (Berkeley, Calif.: Scholar Press, 1983), 138.

2. Elizabeth N. Armstrong, "Hercules and the Muses: Public Art at the Fair," in *The Anthropology of World's Fairs*, 123–24.

3. Gail Bederman, *Manliness and Civilization: A Cultural History of Gender and Race in the United States, 1880–1917* (Chicago: University of Chicago Press, 1995), 31; see also Robert Rydell, *World of Fairs: The Century-of-Progress Expositions* (Chicago: University of Chicago Press, 1993).

4. Charlotte Perkins Gilman, *Women and Economics* (Boston: Small, Maynard and Co., 1899), xxxix.

5. T. J. Jackson Lears, *No Place of Grace: Antimodernism and the Transformation of American Culture, 1880–1920* (New York: Pantheon Books, 1981), 108; Bederman, *Manliness and Civilization,* 12; Peter Filene, *Him/Her/Self: Sex Roles in Modern America,* 2d ed. (Baltimore: Johns Hopkins University Press, 1986), 73.

6. Bederman, *Manliness and Civilization,* 14; Alan Trachtenberg, *The Incorporation of America: Culture and Society in the Gilded Age* (New York: Hill & Wang, 1982), 7; see also Olivier Zunz, *Making America Corporate 1870–1920* (Chicago: University of Chicago Press, 1990), 11–36.

7. George M. Beard, *American Nervousness: Its Causes and Consequences* (New York: Putnam, 1881), 3, quoted in Bederman, *Manliness and Civilization,* 85.

8. John G. Cawelti, *Apostles of the Self-Made Man* (Chicago: University of Chicago Press, 1965), 153.

9. Bederman, *Manliness and Civilization,* 14; see also Lears, *No Place of Grace.*

10. In 1892, the violence of lynching reached its peak, with 161 African Americans murdered by white mobs. Bederman, *Manliness and Civilization,* 47; Jacquelyn Dowd Hall, *Revolt against Chivalry: Jessie Daniel Ames and the Women's Campaign against Lynching* (New York: Columbia University Press, 1974).

11. See Elliott J. Gorn, *The Manly Art: Bare-Knuckle Prize Fighting in America* (Ithaca, N.Y.: Cornell University Press, 1986).

12. John Higham, *Strangers in the Land: Patterns of American Nativism 1860–1925* (New York: Atheneum, 1971).

13. Roger Daniels, *Coming to America: A History of Immigration and Ethnicity in American Life* (New York: HarperCollins, 1990), 125.

14. Joanne J. Meyerowitz, *Women Adrift: Independent Wage Earners in Chicago, 1880–1930* (Chicago: University of Chicago Press, 1988), xvii.

15. Mary E. Odem, *Delinquent Daughters: Protecting and Policing Adolescent Female Sexuality in the United States, 1885–1920* (Chapel Hill: University of North Carolina Press, 1995), 1.

16. Kathy Peiss, "'Charity Girls' and City Pleasures: Historical Notes on Working-Class Sexuality, 1880–1920," in *Passion and Power: Sexuality in History,* ed. Christina Simmons and Kathy Peiss (Philadelphia: Temple University Press, 1989).

17. Carroll Smith-Rosenberg, "Discourses of Sexuality and Subjectivity: The New Woman, 1870–1936," in *Hidden from History: Reclaiming the Gay and Lesbian Past,* ed. Martin Duberman, Martha Vicinus, and George Chauncey Jr. (New York: New American Library, 1989), 265; Filene, *Him/Her/Self,* 26, 238; Barbara Welter, "The Cult of True Womanhood, 1820–1860," *American Quarterly* 18 (1966): 131–75.

18. Anna A. Rogers, "Why American Marriages Fail," *Atlantic Monthly* 100 (September 1907): 289–98, quoted in Filene, *Him/Her/Self,* 42.

19. Filene, *Him/Her/Self,* 144, 42.

20. James Reed, *From Private Vice to Public Virtue: The Birth Control Movement and American Society since 1830* (Princeton, N.J.: Princeton University Press, 1978), 4; Allan Brandt, *No Magic Bullet: A Social History of Venereal Disease in the United States since 1880* (New York: Oxford University Press, 1987), 7; D'Emilio and Freedman, *Intimate Matters,* 173–74.

21. Diane Paul, *Controlling Human Heredity, 1865 to the Present* (Atlantic Highlands, N.J.: Humanities Press, 1995), 102.

22. Theodore Roosevelt, "National Life and Character" (1894) in *American Ideals and Other Essays, Social and Political* (New York: Putnam, 1897), 117, quoted in Bederman, *Manliness and Civilization,* 201.

23. Bederman, *Manliness and Civilization,* 202; "The Question of the Birth Rate," *Popular Science Monthly* 62 (April 1903): 567, quoted in Bederman, *Manliness and Civilization,* 202.

24. Reed, *From Private Vice to Public Virtue,* 201–2.

25. Paula Fass, *The Damned and the Beautiful: American Youth in the 1920s* (New York: Oxford University Press, 1977), 393; D'Emilio and Freedman, *Intimate Matters,* 174.

26. Dorothy Dunbar Bromley, "Minutes of Round Table Discussion, 'The Eugenic Effect of Contraception—the Significance of the Decline in the Birth Rate,'" 13 May 1937, *National Committee on Maternal Health Records,* quoted in Reed, *From Private Vice to Public Virtue,* 211–12.

27. Dr. Alfred Mjoen, quoted in "Scientist Says Modern Women Ruining Race," *Watsonville Pajaronian,* 3 March 1930, p. 1.

28. Bederman, *Manliness and Civilization,* 26; Ronald G. Walters, *Primers for Prudery: Sexual Advice to Victorian America* (Englewood Cliffs, N.J.: Prentice-Hall, 1974), 15.

29. Gilman, *Women and Economics,* 207.

30. Daniel Rogers, "In Search of Progressivism," *Reviews in American History* 10 (December 1982): 113–32; William Deverell and Tom Sitton, eds., *California Progressivism Revisited* (Berkeley: University of California Press, 1994); Odem, *Delinquent Daughters,* 99–100; Alan Dawley, *Struggles for Justice: Social Responsibility and the Liberal State* (Cambridge, Mass.: Harvard University Press 1991), 98–105.

31. Dawley, *Struggles for Justice,* 99; see also Robert Wiebe, *The Search for Order, 1877–1920* (New York: Hill & Wang, 1967).

32. For a thorough analysis of eugenic influence on American immigration policy, see Kevles, *In the Name of Eugenics,* ch. 7.

33. Paul, *Controlling Human Heredity,* 4; Francis Galton, *Inquiries into the Human Faculty* (Macmillan, 1883), 24–25, in Kevles, *In the Name of Eugenics,* ix.

34. Paul, *Controlling Human Heredity,* 3–5.

35. See, for example, Donald K. Pickens, *Eugenics and the Progressives* (Nashville, Tenn.: Vanderbilt University Press, 1968); Haller, *Eugenics;* Hamilton Cravens, *The Triumph of Evolution: American Scientists and the Heredity-Environment Controversy* (Philadelphia: University of Pennsylvania Press, 1978); Degler, *In Search of Human Nature.* For a solid critique of the overemphasis on institutional eugenics in eugenics historiography, see Dikötter, "Race Culture," 467–75; Stepan, *"The Hour of Eugenics,"* 6.

36. Dr. Alfred Mjoen, quoted in "Scientist Says Modern Women Ruining Race," 1.

37. Rydell, *World of Fairs*, 39.

38. Calvin Brown, "Presentation of Medal," in *Official Proceedings of the Second National Conference on Race Betterment* (Battle Creek, Mich.: Race Betterment Foundation [1916?]), 6.

39. Brown, *Official Proceedings*, 4, 5.

40. Lothrop Stoddard, *The Revolt against Civilization* (New York: Scribner, 1923), 94.

41. For a complete history of mental retardation and "feeblemindedness," see James W. Trent, *Inventing the Feeble Mind: A History of Mental Retardation in the United States* (Berkeley: University of California Press, 1994).

42. Stoddard, *The Revolt against Civilization*, 94.

43. Charles Darwin, *The Descent of Man*, (London: J. Murray, 1871), 177, quoted in Paul, *Controlling Human Heredity*, 34.

44. Brown, *Official Proceedings*, 138.

45. Ibid., 140.

46. Ibid., 143.

47. See Peiss, "'Charity Girls'"; Odem, *Delinquent Daughters;* D'Emilio and Freedman, *Intimate Matters*.

48. Brown, *Official Proceedings*, 140.

49. Charlotte Perkins Gilman, "Progress through Birth Control," *North American Review* 224 (December 1927): 627–28.

50. Even feminists expressed concern over how to reconcile individual rights and freedoms with their role as wives and mothers. As Cott notes, feminists in the 1920s emphasized equality over difference, downplayed female distinctiveness, and thereby lost much of the momentum and unity they had found in the suffrage movement; Nancy F. Cott, *The Grounding of Modern Feminism* (New Haven, Conn.: Yale University Press, 1987).

51. Hinkle, "Women and the New Morality," 248.

52. "California Civic League Meets in Convention," *California Women's Bulletin* 3, no. 3 (January 1915): 27.

53. Elizabeth Lunbeck, "'A New Generation of Women': Progressive Psychiatrists and the Hypersexual Female," *Feminist Studies* 13, no. 3 (Fall 1987): 513. Lunbeck argues in *The Psychiatric Persuasion: Knowledge, Gender, and Power in Modern America* (Princeton, N.J.: Princeton University Press, 1994) that psychiatric knowledge and power resulted in an "epistemological transformation that relegated psychiatrists to a status as cultural arbiters of sexual modernity" in the early twentieth century (5). Yet other studies challenge this argument by citing dramatically different institutional and reform experiences taking place simultaneously in different parts of the country. As Regina Kunzel notes in *Fallen Women, Problem Girls: Unmarried Mothers and the Professionalization of Social Work, 1890–1945* (New Haven, Conn.: Yale University Press, 1993), "Work with unmarried mothers, apparently much like social work, was largely unaffected by psychiatric ideas through the 1920s and 1930s" (148). Psychiatry alone cannot be held responsible for a major shift in the perception of sexuality; other social movements, including eugenics, contributed to this shift.

54. For more on Progressive-era concerns about female sexual delinquency, see Ruth M. Alexander, *The "Girl Problem": Female Sexual Delinquency in New York, 1900–1930* (Ithaca, N.Y.: Cornell University Press, 1995), and Odem, *Delinquent Daughters.*

55. D'Emilio and Freedman, *Intimate Matters,* 171–201.

56. Pernick, *The Black Stork,* 15.

57. Paul, *Controlling Human Heredity,* 46.

58. See Kevles, *In the Name of Eugenics,* 41–56.

59. See Leila Zenderland, "The Debate over Diagnosis: Henry Herbert Goddard and the Medical Acceptance of Intelligence Testing," in *Psychological Testing and American Society, 1880–1930,* ed. Michael M. Sokal (New Brunswick, N.J.: Rutgers University Press, 1987). She offers a useful analysis of the debate between physicians and psychologists over the most effective procedure for understanding and treating feeblemindedness in state institutions.

60. Ibid., 65.

61. Ibid., 62.

62. JoAnne Brown, *The Definition of a Profession: The Authority of Metaphor in the History of Intelligence Testing, 1890–1930* (Princeton, N.J.: Princeton University Press, 1992), 41.

63. Walter E. Fernald, *The Burden of Feeble-mindedness* (Boston: Massachusetts Society for Mental Hygiene, 1918), 6.

64. See Douglas Baynton, *Forbidden Signs: American Culture and the Campaign against Sign Language* (Chicago: University of Chicago Press, 1996).

65. Brown, *The Definition of a Profession,* 39.

66. Goddard was also pointing out that "feeblemindedness" had been used inconsistently; sometimes it referred to what he called "the entire range of mental defectives" and sometimes specifically to the highest grade. Henry H. Goddard, "Four Hundred Feeble-minded Children Classified by the Binet Method," *Journal of Genetic Psychology* 17, no. 3 (1910): 445. Emphasis added.

67. Ibid. Emphasis added.

68. Baynton, *Forbidden Signs.*

69. Goddard, "Four Hundred Feeble-minded Children," 445.

70. Bederman, *Manliness and Civilization,* 88.

71. Ibid., 92; Degler, *In Search of Human Nature,* 4–47; Dorothy Ross, *G. Stanley Hall: The Psychologist as Prophet* (Chicago: University of Chicago Press, 1972), 90.

72. Bederman, *Manliness and Civilization,* 97.

73. Ibid., 94.

74. Paul, *Controlling Human Heredity,* 40–49. Stepan and Pernick both demonstrate, however, that until 1915 in the United States and well into the 1940s in Latin America, neo-Lamarckism retained scientific credibility in many professional circles. Stepan, *"In the Hour of Eugenics,"* 194–95; Pernick, *The Black Stork,* 42–48.

75. Goddard was not the first scientist to link social deviance with degeneracy. Nineteenth-century Italian criminal anthropologist Cesare Lombroso believed that the criminal was an "atavism," linked to less evolved forms of human life. Nineteenth-century German psychiatrist Richard von Krafft-Ebing believed that

homosexuals were trapped in a more primitive stage of evolutionary development. See David G. Horn, "This Norm Which Is Not One: Reading the Female Body in Lombroso's Anthropology," and Jennifer Terry, "Anxious Slippages between 'Us' and 'Them': A Brief History of the Scientific Search for Homosexual Bodies," both in *Deviant Bodies: Critical Perspectives on Difference in Science and Popular Culture,* ed. Jennifer Terry and Jacqueline Urla (Bloomington: Indiana University Press, 1995), 112, 135.

76. For a useful analysis of the significance of Goddard's Kallikak study, see Trent, *Inventing the Feeble Mind,* 163–65.

77. Contributors to the anthology *Deviant Bodies* offer additional evidence that various "deviant bodies," including the homosexual body, the HIV-infected body, and the criminal body, "have been used implicitly and intricately to shore up notions of what is normal and what is not." Jacqueline Urla and Jennifer Terry, "Introduction: Mapping Embodied Deviance," in *Deviant Bodies: Critical Perspectives on Difference in Science and Popular Culture,* ed. Jennifer Terry and Jacqueline Urla (Bloomington: Indiana University Press, 1995), 4–5. Sander Gilman, *Difference and Pathology: Stereotypes of Sexuality, Race, and Madness* (Ithaca, N.Y.: Cornell University Press, 1985).

78. Goddard, "Four Hundred Feeble-minded Children," 447.

79. Ibid. Emphasis added.

80. Ibid. Because Goddard had just suggested using the term *moron* to distinguish the highest grade of feeblemindedness, he was not yet using it in his analysis; instead he refers to this class as the "moral imbecile," Isaac Kerlin's (superintendent of the Pennsylvania asylum in the 1850s) term. In later works, he used *moron.* Other eugenicists continued to use the terms *moral imbeciles* and *moron* haphazardly.

81. Henry H. Goddard, *The Menace of Mental Deficiency from the Standpoint of Heredity* (Vineland: New Jersey Training School, 1915), 9.

82. "California Civic League Meets in Convention," 27.

83. Massachusetts League for Preventive Work, *Feeble-minded Adrift* (Boston: n.p., 1916), 1. The League was formed in 1916 by the Massachusetts Society for the Prevention of Cruelty to Children, one of twenty organizations (including the Boston Children's Aid Society and the Milk and Baby Hygiene Association) that made up the larger League. See Linda Gordon, *Heroes of Their Own Lives: The Politics and History of Family Violence* (New York: Penguin Books, 1988), 76.

84. Massachusetts League for Preventive Work, *Feeble-Minded Adrift,* 1–13.

85. Paul Popenoe to Charles Davenport, 22 October 1913, Popenoe file, Charles Davenport Papers, American Philosophical Society, Philadelphia.

86. Gordon, *Heroes,* 73. For other works on the Progressive era that emphasize a generational shift in female moral reformers, see Odem, *Delinquent Daughters;* Kunzel, *Fallen Women, Problem Girls.*

87. Gordon, *Heroes,* 73.

88. Odem, *Delinquent Daughters,* 105.

89. Fernald, *The Burden of Feeble-mindedness,* 10.

90. Irving Fisher to Lewis Terman, 12 June 1922, Terman Papers, SC 38, 2–5.

91. Ann J. Lane, introduction to *Herland,* by Charlotte Perkins Gilman (New York: Pantheon, 1979; *Herland* originally published 1915), xiv.

92. Molly Ladd-Taylor, *Mother Work: Women, Child Welfare, and the State* (Urbana: University of Illinois Press, 1994), 4; Filene, *Him/Her/Self,* 43–44.

93. Deverell and Sitton, *California Progressivism Revisited,* 3. Dowbiggin also notes that the influence of Progressivism was felt "demonstrably" in California; Dowbiggin, *Keeping America Sane,* 125.

94. Larson, *Sex, Race, and Science,* 38.

Chapter Two

1. For more on eugenics exhibitions at state and world's fairs, see Robert Rydell, "'Fitter Families for Future Firesides': Eugenics Exhibitions between the Wars," in *World of Fairs,* 38–58.

2. Fred Butler, "The Importance of Out-Patient Clinics in State Institutions," *American Journal of Mental Deficiency* 45 (1940): 78.

3. E. A. Whitney, "Introduction of the President," *American Journal of Mental Deficiency* 47 (1942): 5.

4. For more on birth control and eugenics, see chapter 3.

5. Whitney, "Introduction of the President," 5.

6. California Home for the Care and Training of Feeble-Minded Children, *Circular of Information* (Sacramento: California State Printing Office, 1887), 10.

7. State Board of Charities and Corrections, *Biennial Report* (Sacramento: State Printing Office, 1922), 49.

8. Robert C. Elliot, "20,000 Minds in Retreat: The Story of Our Mental Crackups," *San Francisco News,* 20 March 1936, p. 23.

9. California Home, *Circular of Information,* 10.

10. Esther Pond and Stuart Brody, *Evolution of Treatment Methods at a Hospital for the Mentally Retarded,* California Mental Health Research Monograph 3 (Sacramento: California Department of Mental Hygiene, 1965), 1.

11. State Commission in Lunacy, *Biennial Report* (Sacramento: State Printing Office, 1904), 11.

12. Trent, *Inventing the Feeble Mind,* 100.

13. California Home, *Circular of Information,* 10.

14. Ibid.

15. David J. Rothman, *The Discovery of the Asylum: Social Order and Disorder in the New Republic* (Boston: Little, Brown, 1971), 151.

16. California Home, *Circular of Information,* 10.

17. Steven Mintz and Susan Kellogg, *Domestic Revolutions: A Social History of American Family Life* (New York: Free Press, 1988), 58.

18. California Home, *Circular of Information,* 10.

19. State Commission in Lunacy, *Biennial Report* (1904), 102–3.

20. Trent, *Inventing the Feeble Mind,* 141.

21. State Commission in Lunacy, *Biennial Report* (1904), 12.

22. State Commission in Lunacy, *Biennial Report* (1908), 20.

23. State Board of Charities and Corrections, *Biennial Report* (1908), 73–74.

24. Odem, *Delinquent Daughters,* 136.

25. State Board of Charities and Corrections, *Biennial Report* (1908), 73–74.

26. State Board of Charities and Corrections *Biennial Report* (1914), 112.

27. See, for example, Brown, *The Definition of a Profession;* Baynton, *Forbidden Signs.*

28. State Commission in Lunacy, *Biennial Report* (1914), 15.

29. Henry L. Minton, *Lewis M. Terman: Pioneer in Psychological Testing* (New York: New York University Press, 1988), 48–49.

30. Quoted in Paul Davis Chapman, *Schools as Sorters: Lewis M. Terman, Applied Psychology, and the Intelligence Testing Movement, 1890–1930* (New York: New York University Press, 1988), 32. Though Terman claims to be referring to "every feeble-minded individual," his emphasis is clearly specifically on women, as he refers to their "reproductive period."

31. State Board of Charities and Corrections, *Biennial Report* (1914), 115; Lewis Terman, *The Measurement of Intelligence* (Boston: Houghton Mifflin, 1916), 111, quoted in Stephen Jay Gould, *The Mismeasure of Man* (New York: Norton, 1981), 181; Lewis Terman, "The Measurement of Intelligence," in Russell Jacoby and Naomi Glauberman, *The Bell Curve Debate: History, Documents, Opinions* (New York: Random House, 1995), 545.

32. State Commission in Lunacy, *Biennial Report* (1914), 18. Emphasis added.

33. State Board of Charities and Corrections, *Biennial Report* (1914), 115–16.

34. State Board of Charities and Corrections, *Biennial Report* (1916), 32.

35. Trent, *Inventing the Feeble Mind,* 89.

36. State Board of Charities and Corrections, *Biennial Report* (1916), 32, 31. In his study of southern institutions for the mentally retarded during the first four decades of the twentieth century, Noll also finds that many female patients were labeled "morons" simply because of "excessive sexual activity," with no other verification of intellectual disability. Steven Noll, *Feeble-minded in Our Midst: Institutions for the Mentally Retarded in the South, 1900–1940* (Chapel Hill: University of North Carolina Press, 1995), 116. Kevles also notes the circular argument made by eugenicists, who claimed that immoral behavior was evidence of feeblemindedness, which in turn resulted in immoral behavior. See Kevles, *In the Name of Eugenics,* 107.

37. California State Board of Charities and Corrections, *Surveys in Mental Deviation* (Sacramento: California State Printing Office, 1918), 41, 51.

38. Ibid., 51. Emphasis added.

39. Ibid., 62. Horn notes a similar trend in nineteenth-century Italy, where criminologist Cesare Lombroso's attempt to strengthen the boundaries between the "female offender" and "normal" women instead revealed how unstable these boundaries were. Horn, "This Norm Which Is Not One," 109.

40. California State Board of Charities and Corrections, *Surveys in Mental Deviation,* 85, 62, 65, 66.

41. Ibid., 66.

42. D'Emilio and Freedman, *Intimate Matters,* 201; Alexander, *The "Girl Problem,"* 59.

43. Judith Walzer Leavitt, *Typhoid Mary: Captive to the Public's Health* (Boston: Beacon Press, 1996), 23–26; James H. Cassedy, *Medicine in America* (Baltimore: Johns Hopkins University Press, 1991), 110.

44. Brown, *The Definition of a Profession,* 81.

45. Walter E. Fernald, "Care of the Feeble-minded," in *Proceedings, National Conference of Charities and Corrections* (Indianapolis: Press of Frederick J. Heer, 1904), 383; Fernald, *The Burden of Feeble-mindedness,* 10.

46. Nancy Rockafellar, "Making the World Safe for the Soldiers of Democracy: Patriotism, Public Health and Venereal Disease Control on the West Coast, 1910–1919" (Ph.D. diss., University of Washington, 1990), 32, 431. See also Brandt, *No Magic Bullet;* Ruth Rosen, *The Lost Sisterhood: Prostitution in America, 1900–1918* (Baltimore: Johns Hopkins University Press, 1982).

47. Prince Morrow, *Social Diseases and Marriage* (New York, 1904), 78–79, quoted in Brandt, *No Magic Bullet,* 14. Brandt points out the rudimentary understanding of "heredity" in the early twentieth century on the part of physicians (as syphilis was not a genetic disorder). But as late as 1939 physicians still differentiated between "hereditary syphilis" and "acquired syphilis." In *The Venereal Peril,* Dr. William L. Holt targeted venereal diseases as "agents in race suicide and degeneration" (New York: Eugenics Publishing, 1939), 7.

48. State Commission in Lunacy, *Biennial Report* (1918), 15.

49. Rockafellar, "Making the World Safe," 402, 206–7; State Commission in Lunacy, *Biennial Report* (1912), 77. Brown writes that psychiatrists commonly compared mental tests to the Wassermann test for syphilis, information she obtained through personal communication with historian Elizabeth Lunbeck; Brown, *The Definition of a Profession,* 91.

50. See Odem, *Delinquent Daughters,* for a historical analysis of the change from social-purity reform to progressive social work between 1885 and 1920.

51. Brandt, *No Magic Bullet,* 31; Mark Connelly, *The Response to Prostitution in the Progressive Era* (Chapel Hill: University of North Carolina Press, 1980), 42–43; Dr. A. J. McLaughlin, "Pioneering in Venereal Disease Control," *American Journal of Obstetrics* 80 (December 1919): 639, quoted in Brandt, *No Magic Bullet,* 92. Both Brandt and Rosen note the early-twentieth-century assumption on the part of doctors that prostitutes were feebleminded, but they do not discuss the relationship between "feeblemindedness" and eugenics. As a result, neither emphasizes the link between female sexuality and racial degeneracy. Brandt, *No Magic Bullet,* 92; Rosen, *The Lost Sisterhood,* 22.

52. Rosen, *The Lost Sisterhood,* 28–33; Rockafellar, "Making the World Safe," 131.

53. Bascom Johnson, "Next Steps," *Journal of Social Hygiene* 4(1918): 9.

54. Odem, *Delinquent Daughters,* 125–26.

55. *San Francisco Chronicle,* 15 February 1918; 4 May 1918.

56. State Commission in Lunacy, *Biennial Report* (1918), 75–76.

57. *San Francisco Chronicle,*19 March 1918, p. 5.

58. Fernald, *The Burden of Feeble-mindedness,* 10.

59. Odem, *Delinquent Daughters,* 122; Alexander, *The "Girl Problem,"* 21; Rockafellar, "Making the World Safe," 245.

60. California State Board of Charities and Corrections, *Surveys in Mental Deviation,* 66.

61. Leavenworth, "Virtue for Women," 102.

62. J. H. Landman, *Human Sterilization: The History of the Sexual Sterilization Movement* (New York: Macmillan, 1932), 14.

63. State Commission in Lunacy, *Biennial Report* (1914), 15.

64. State Commission in Lunacy, *Biennial Report* (1912), 18.

65. Landman, *Human Sterilization,* 58.

66. State Commission in Lunacy, *Biennial Report* (1916), 33. The term *asexualization* was later replaced with *sterilization,* as asexualization was associated with castration and the increasingly unpopular notion of "unsexing."

67. Department of Institutions, *Biennial Report* (Sacramento: State Printing Office, 1926), 92. Emphasis added.

68. See chapter 1.

69. Dowbiggin, *Keeping America Sane,* 121.

70. State Commission in Lunacy, *Biennial Report* (1914), 13–14.

71. Medical professionals expended a great deal of effort distinguishing between the feebleminded and the insane, as etiology and treatment of the two disorders differed significantly. Richard Fox writes in *So Far Disordered in Mind: Insanity in California, 1870–1930* (Berkeley: University of California Press, 1978) that "in the strictest sense the 'feebleminded' were distinct from the 'insane,' but in both popular and professional usage there was considerable overlapping between the two" (32). Admission to Sonoma required medical proof that the subject was not insane "by at least one physician of actual practice and recognized by the State Board of Medical Examiners"; California Home, *Circular of Information,* 7.

72. State Commission in Lunacy, *Biennial Report* (1914), 95.

73. State Board of Charities and Corrections, *Biennial Report* (1918), 53.

74. Phil Esler, "Sterilization of Mental Patients? Some Doctors Say Yes, Some No," *Santa Rosa Press Democrat,* 15 October 1961, in AVS box 14, Butler file, Association for Voluntary Sterilization Papers, Social Welfare History Archives, University of Minnesota, Minneapolis.

75. Whitney, "Introduction of the President," 5; Robert Dickinson to Curtis Wood, December 1948, AVS box 2 folder 15.

76. Butler, "The Importance of Out-Patient Clinics in State Institutions," 78.

77. Butler to Dr. Karl M. Bowman, 11 January 1951, AVS box 4 folder 34.

78. See Brown, *The Definition of a Profession,* for ways in which early-twentieth-century psychologists used medical language to gain authority.

79. Butler, "The Importance of Out-Patient Clinics in State Institutions," 78.

80. State Commission in Lunacy, *Biennial Report* (1920), 59.

81. In both private correspondence and published works, Butler discussed his role in the sterilization campaign into the 1960s. See, for example, Esler, "Sterilization of Mental Patients?"

82. Fred Butler, "Selective Sterilization: Discussion," *Journal of Psycho-Asthenics* 35 (1930): 65.

83. F. O. Butler, "Sterilization Procedure and Its Success in California Institutions" (address given at the National Conference of Juvenile Agencies, 1925), 3.

84. State Board of Charities and Corrections, *Biennial Report* (1918–20); State Commission in Lunacy, *Biennial Report* (1918–20); Department of Institutions, *Biennial Report* (1922–32); Department of Institutions, *Statistical Report* (Sacramento: State Printing Office, 1934–45); F. O. Butler and Clarence J. Gamble, "Sterilization in a California School for the Mentally Deficient," *American Journal of Mental Deficiency* 6 (April 1947), 745–47.

85. Materials used for analysis are from two different sources. The first source is 249 records of patients sterilized at Sonoma between 1922 and 1925. Though they were initially compiled by Sonoma staff, the surviving records are

copies compiled by the Human Betterment Foundation, a nonprofit eugenics organization founded to research and publicize the results of California's eugenic-sterilization program; these records are now part of the Gosney Papers at the California Institute of Technology Archives. The drawback is that these records are not inclusive. Those patients who were not sterilized at Sonoma—because they either were not of childbearing age or were not candidates for release from the institution because of the extent of their disability—remain hidden from view. There is thus no opportunity for comparison in order to establish criteria for sterilization selection. However, sociologist Judith Grether found that 93 percent of all cases were sterilized at Sonoma between 1922 and 1934; Judith K. Grether, "Sterilization and Eugenics: An Examination of Early Twentieth Century Population Control in the United States" (Ph.D. diss., University of Oregon, 1980), 95. The second source is patient records from the Sonoma State Home, which are inclusive and contain more detail but end in August 1919. I analyzed 164 records dating from January 1918 to August 1919. These are currently housed in the California State Archives in Sacramento. The notes here will specify which collection specific statistical information comes from.

86. Department of Institutions, *Biennial Report* (1922), 80.

87. Paul Popenoe to Gosney, 25 March 1926, E. S. Gosney Papers and the Records of the Human Betterment Foundation, 7.2, Department of Special Collections, Stanford, Calif.

88. Sonoma State Home Sterilization Records, 1922–25, Gosney Papers.

89. Sonoma State Home Patient Records, 1918–19, California State Archives; Terry, "Anxious Slippages between 'Us' and 'Them,'" 141. Terry's essay offers a thorough investigation and analysis of Robert Dickinson's influence on the Sex Variant study (conducted between 1935 and 1941 by the Committee for the Study of Sex Variants) as well as on medical attitudes toward homosexuality and the search for physical differences between the "normal" and the "sex-variant" body.

90. Sonoma State Home Sterilization Records, Gosney Papers; George Chauncey Jr., "From Sexual Inversion to Homosexuality: The Changing Medical Conceptualization of Female Deviance," in *Passion and Power: Sexuality in History,* ed. Christina Simmons and Kathy Peiss (Philadelphia: Temple University Press, 1989), 90; Lewis Terman and Catherine Cox Miles, *Sex and Personality: Studies in Masculinity and Femininity* (New York: Russell & Russell, 1936), 256, 248; Terry, "Anxious Slippages between 'Us' and 'Them,'" 135.

91. Paul Popenoe, "Success on Parole after Sterilization," reprinted from the *Proceedings of the Fifty-First Annual Session of the American Association for the Study of the Feebleminded* (June 1927), in *Collected Papers on Eugenic Sterilization in California: A Critical Study of Results in 6000 Cases* (Pasadena: Human Betterment Foundation, 1930), 5.

92. Paul Popenoe, "Eugenic Sterilization in California: The Feebleminded," reprinted from the *Journal of Social Hygiene* 13, no. 5 (May 1927), in *Collected Papers on Eugenic Sterilization in California: A Critical Study of Results in 6000 Cases* (Pasadena: Human Betterment Foundation, 1930), 328. Emphasis added.

93. E. S. Gosney and Paul Popenoe, *Twenty-Eight Years of Sterilization in California* (Pasadena, Calif.: Human Betterment Foundation, 1938), 33.

94. E. S. Gosney and Paul Popenoe, *Sterilization for Human Betterment: A Summary of Results of 6,000 Operations in California, 1909–1929* (New York: Macmillan, 1929), 100.

95. Sonoma State Home Sterilization Records, Gosney Papers.

96. Ibid.

97. Ibid.

98. Ibid.

99. Odem, *Delinquent Daughters,* 158.

100. Sonoma State Home Sterilization Records, Gosney Papers

101. In her study of adolescent female sexuality, Odem notes the important role of working-class parents in policing their daughters' behavior. She concludes that "even as the state assumed greater authority over moral behavior, families still continued to play an important role in the regulation of sexual conduct." Odem, *Delinquent Daughters,* 184.

102. Sonoma State Home Sterilization Records, Gosney Papers.

103. Fred Butler, "The Prevention of Mental Deficiency by Sterilization, 1949," unpublished manuscript, AVS box 14 folder 113. In Popenoe's research on California sterilization, he did not consider the testimony of patients at Sonoma to be relevant. "We made no attempt to get direct expressions of opinion from those sterilized at the state home for the feebleminded, believing that their testimony would not be valuable, in view of their mental levels." Popenoe, "Eugenic Sterilization in California: The Feebleminded," 283.

104. Popenoe, "Eugenic Sterilization in California: The Feebleminded," 324.

105. Helen Montague, "The Causes of Delinquency in Mentally Defective Boys," *Journal of Psycho-Asthenics* 35 (1930): 104.

106. Department of Institutions, *Biennial Report* (1922), 26; Grether, "Sterilization and Eugenics," 160–63. For more on racism and nativism in California, see Roger Daniels, *The Politics of Prejudice: The Anti-Japanese Movement in California and the Struggle for Japanese Exclusion* (Berkeley: University of California Press, 1978); Dowbiggin, *Keeping America Sane,* 120.

107. The patient population at Sonoma and those selected for sterilization do not appear to have been targeted based on their racial or ethnic background. Assuming that race would play a significant role in California's sterilization policy, Grether was surprised to find scant evidence of it in her statistical analysis of sterilization in California. "It was expected that those who were sterilized would be disproportionately of minority racial or ethnic identity and would be disproportionately foreign-born," she writes. "There is very little evidence in the data to support this expectation. According to these data, the foreign-born were only very slightly more likely to be sterilized than was true for the native-born." In regard to race, she found "no racial differences between the sterilized and unsterilized." Grether, "Sterilization and Eugenics," 129.

108. Sonoma State Home Sterilization Records, Gosney Papers. The social worker quoted did not specify the race of the patient to whom she was referring.

109. "California Civic League Meets in Convention," *California Woman's Bulletin* 3, no. 3 (January 1915): 27.

110. State Commission in Lunacy, *Biennial Report* (1916), 33.

111. State Commission in Lunacy, *Biennial Report* (1918), 77.

Chapter Three

1. D'Emilio and Freedman, *Intimate Matters*, 241.

2. See Walters, *Primers for Prudery*.

3. Nancy F. Cott, "Passionlessness: An Interpretation of Victorian Sexual Ideology, 1790–1850," in *A Heritage of Her Own*, ed. Nancy F. Cott and Elizabeth H. Pleck (New York: Simon & Schuster, 1979).

4. Gordon, *Woman's Body, Woman's Right*, 175.

5. Cynthia Eagle Russett, *Sexual Science: The Victorian Construction of Womanhood* (Cambridge, Mass.: Harvard University Press, 1989), 104–29.

6. Dr. Sylvester Graham, *Chastity, in a Course of Lectures to Young Men: Intended Also for the Serious Consideration of Parents and Guardians* (New York: Fowler and Wells, n.d.), 15, quoted in Walters, *Primers for Prudery*, 34. Emphasis in original.

7. Ellen Chesler, *Woman of Valor: Margaret Sanger and the Birth Control Movement in America* (New York: Doubleday, 1992), 123; Paul Robinson, *The Modernization of Sex: Havelock Ellis, Alfred Kinsey, William Masters and Virginia Johnson* (New York: Harper & Row, 1976), 2.

8. Havelock Ellis, *Psychology of Sex* (London: Heinemann, 1946), 3, quoted in Jeffrey Weeks, *Sexuality and Its Discontents: Meanings, Myths & Modern Sexualities* (London: Routledge, 1985), 62.

9. Havelock Ellis, "The Meaning of Purity," in *Little Essays of Love and Virtue* (New York, 1922), 60–61, quoted in Robinson, *The Modernization of Sex*, 28.

10. Frederick Lewis Allen, *Only Yesterday: An Informal History of the 1920s* (New York: Harper & Row, 1964), 96.

11. Katharine Bement Davis, *Factors in the Sex Life of Twenty-Two Hundred Women* (New York: Harper, 1929), 14; D'Emilio and Freedman, *Intimate Matters*, 175, 174.

12. Elaine May, *Barren in the Promised Land: Childless Americans and the Pursuit of Happiness* (New York: Basic Books, 1995), 87.

13. Richard A. Soloway, "The 'Perfect Contraceptive': Eugenics and Birth Control Research in Britain and America in the Interwar Years," *Journal of Contemporary History* 30 (1995): 639.

14. Christopher Lasch, *Haven in a Heartless World: The Family Besieged* (New York: Norton, 1977), 8–9.

15. Reed, *From Private Vice to Public Virtue*, 68.

16. See Gordon, *Woman's Body, Woman's Right*, 255; Reed, *From Private Vice to Public Virtue*, 34–68, 104–5, 113, 131–32, 144.

17. G. Stanley Hall to Mary Lawrence East, 13 December 1916, Blanche Ames Papers, Sophia Smith Collection, Smith College, quoted in Reed, *From Private Vice to Public Virtue*, 104–5. For more on women reformers shunning birth control, see Reed, *From Private Vice to Public Virtue*, 131–32.

18. Reed, *From Private Vice to Public Virtue*, 144; Chesler, *Woman of Valor*, 269; Carole R. McCann, *Birth Control Politics in the United States, 1916–1945* (Ithaca, N.Y.: Cornell University Press, 1994), 78.

19. McCann, *Birth Control Politics*, 26, 51, 40.

20. Gordon, *Woman's Body, Woman's Right*, 275; McCann, *Birth Control Politics*, 20, 78; Chesler, *Woman of Valor*, 196.

21. Lydia DeVilbiss to Human Betterment Foundation, 8 May 1941, Gosney Papers, 7.6; Margaret Sanger, "Why Not Birth Control in America?" *Birth Control Review,* May 1919, 10–11, quoted in Gordon, *Woman's Body, Woman's Right,* 277; Gordon, *Woman's Body, Woman's Right,* 278.

22. Raymond Pearl to Robert Dickinson, 27 May 1925, Dickinson File, Raymond Pearl Papers, American Philosophical Society, Philadelphia; Harry Laughlin, "Further Studies on the Historical and Legal Development of Eugenical Sterilization in the United States," *Journal of Psycho-Asthenics* 41 (1936): 98.

23. Roswell Johnson to Charles Davenport, 2 December 1921, Johnson File, Davenport Papers.

24. Charles Davenport to Robert Dickinson, 26 July 1923, Dickinson File, Davenport Papers; Charles Davenport to Margaret Sanger, 21 October 1921, Sanger File, Davenport Papers.

25. Charles Davenport to Robert Dickinson, 26 July 1923, Dickinson File, Davenport Papers.

26. Reed, *From Private Vice to Public Virtue,* 165, 156.

27. Robert Dickinson, "Simple Sterilization of Women by Cautery Stricture of Intra-uterine Tubal Openings," *Surgery, Gynecology and Obstetrics* 23 (1916): 185–90, quoted in Reed, *From Private Vice to Public Virtue,* 167; McCann, *Birth Control Politics,* 96; Robert Dickinson, "The Birth Control Movement," abstract of address presented before the Assembly of the Interstate Post-Graduate Medical Association of North America, 18 October 1926, AVS box 15, Dickinson file.

28. Dickinson, "The Birth Control Movement," 7; Robert Dickinson, "The Birth Control Clinic of Today and Tomorrow," *Eugenics* 2 (May 1929): 9–10, quoted in Reed, *From Private Vice to Public Virtue,* 185.

29. Reed, *From Private Vice to Public Virtue,* 182, 158; Chesler, *Woman of Valor,* 270. Most historians of birth control credit Dickinson with greatly influencing the birth-control movement, yet they do not adequately discuss his position in the eugenics movement. See Gordon, *Woman's Body, Woman's Right;* Reed, *From Private Vice to Public Virtue;* Soloway, "The 'Perfect Contraceptive.'"

30. Robert Dickinson, "Sterilization without Unsexing: List of Exhibit Material Shown," unpublished memo, 11 July 1928, AVS box 14 folder 115.

31. E. S. Gosney to Lewis Terman, n.d. [September 1928], Terman Papers, SC 38, 2–11.

32. Regina Morantz-Sanchez, *Conduct Unbecoming a Woman: Medicine on Trial in Turn-of-the-Century Brooklyn* (New York: Oxford University Press, 1999), 107.

33. Reed, *From Private Vice to Public Virtue,* 153.

34. Filene, *Him/Her/Self,* 29, 35, 50.

35. For more on the relationship between nineteenth-century gynecological surgery and eugenic sterilization, see Wendy Kline, "'Building a Better Race': Eugenics and the Making of Modern Morality in America, 1900–1960" (Ph.D. diss., University of California at Davis, 1998); for more on nineteenth-century gynecological surgery, see Morantz-Sanchez, *Conduct Unbecoming a Woman;* Carol Groneman, "Nymphomania: The Historical Construction of Female Sexuality," *Signs,* Winter 1994, 337–67; Andrew Scull and Diane Favreau, "'A Chance to Cut Is a Chance to Cure': Sexual Surgery for Psychosis in Three Nineteenth Century Cities," *Research in Law, Deviance and Social Control* 8 (1986); Elizabeth Sheehan, "Victorian Clitoridectomy: Isaac Baker Brown and His Harmless Operative Pro-

cedure," in *The Gender/Sexuality Reader,* ed. Roger N. Lancaster and Micaela di Leonardo (New York: Routledge, 1997); Ann Dally, *Women under the Knife: A History of Surgery* (New York: Routledge, 1991).

36. Robert Dickinson, "Sterilization without Unsexing: A Surgical Review, with Especial Reference to 5,820 Operations on Insane and Feebleminded in California," reprinted from the *Journal of the American Medical Association* 92 (February 1929), in *Collected Papers on Eugenic Sterilization in California,* 1.

37. Robert Dickinson, letter to the Editor of the *Mirror* (sent also to many other magazines and newspapers), 19 April 1947, AVS box 2 folder 14.

38. Dickinson, "Simple Sterilization," 203.

39. Robert Dickinson, "Sterilization by Adhesions inside Uterine Cornu Following Denudation by Reamer Curette" (paper read before the section on Gynecology and Obstetrics, New York Academy of Medicine, 24 November 1936), AVS box 14 folder 116.

40. Robert Dickinson to Mr. John Emerson, 15 February 1935, AVS box 14 folder 116.

41. Dickinson to Gosney, 20 October 1926, Gosney Papers, 7.2.

42. Gosney to Dickinson, 22 January 1926, Gosney Papers, 7.2.

43. Dickinson to Gosney, 5 February 1926, Gosney Papers, 7.2.

44. Laughlin to Gosney, 7 August 1923, Gosney Papers, 7.2.

45. Laughlin, "Further Studies on the Historical and Legal Development of Eugenical Sterilization," 104. Grossman points out that in Germany during the Weimar Republic doctors generally agreed that women, rather than men, should be the "primary candidates for sterilization, although they acknowledged that male sterilization was considerably simpler and less dangerous; women, after all, were the ones to become pregnant, and even if a husband were sterilized, they could still be raped or seduced." Grossman, *Reforming Sex,* 72–73.

46. Laughlin to Dickinson, 9 August 1927, AVS box 13 folder 106.

47. Dickinson, "memo," 7 June 1928, AVS box 14 folder 115.

48. Dickinson argued that this test, known as tubal insufflation, was "imperative" after every sterilization operation, before parole or discharge. It was used to determine whether the operation had effectively sealed off the fallopian tubes. Dickinson explained his "simple form of insufflation": "A rubber bulb, a T connection, and the glass intra-uterine tube with a shoulder at the external os and multiple opening at the tip constitute an outfit (at the cost of only $2.50) that can be coupled up to any blood pressure gauge.... The speculum exposes the cervix. The tube is passed until its expanded bulb closes the external os snugly. This snugness is clearly observed through the glass.... The bulb is slowly compressed. In open tubes the air goes through normally at from 60 to 80 mm of pressure; in considerably obstructed tubes the pressure ranges from 120 to 150 mm." Dickinson, "Sterilization without Unsexing: A Surgical Review," 18–19.

49. In his early practice, Dickinson made use of domestic servants, who could not pay, as subjects for his research. See Reed, *From Private Vice to Public Virtue,* 155. For more on human experimentation, see Susan Lederer, *Subjected to Science: Human Experimentation in America before the Second World War* (Baltimore: Johns Hopkins University Press, 1995).

50. Dickinson to Laughlin, 15 August 1927, AVS box 13 folder 106.

51. In 1928, there were eight state hospitals in California: two for the "feeble-minded" (Sonoma State Home and Pacific Colony) and six for the insane (at Napa, Stockton, Agnews, Mendocino, Norwalk, and Patton).

52. Dr. W. D. Wagner to all California medical superintendents, 26 February 1926, Gosney Papers, 7.2.

53. Robert Dickinson, "Sterilization without Unsexing: Its Relative Simplicity" (paper read before the Sterilization League of New Jersey, November 12, 1941), AVS box 14 folder 113.

54. "Experiment to be Studied Medically: Sterilization Data Is to Engage Gynecologist," *Pasadena Star News,* n.d. [February 1928], AVS box 15 folder 126.

55. Dickinson, "Sterilization without Unsexing: A Surgical Review," 2.

56. Ibid., 21; Robert Dickinson, "Sterilization of Women: Extended Trial of Cautery Stricture," unpublished memo, 17 August 1927, Gosney Papers, 17.2.

57. Dickinson, "Office Sterilization of Women by Intrauterine Cautery or Reamer at Cornua," 30 October 1935, AVS box 14 folder 116.

58. Dickinson, "Sterilization without Unsexing: A Surgical Review," 23.

59. In a different publication, in which Dickinson analyzed 650 of his female patients, he determined that 10 percent of them were "sufficiently aroused sexually to appear erotic at pelvic examination as shown by congested vulva [and] wetness." Robert Dickinson and Lura Bream, *The Single Woman: A Medical Study in Sex Education* (New York: Reynal & Hitchcock, 1934), xviii.

60. Though Dickinson claimed the procedure was virtually painless, another doctor believed it to be "a great deal more painful" than cauterization of the cervix because he had always noticed the "terrible odor and the awful slough that we get from just a simple cauterization." Dr. K. C. Copenhaver, response to Charles Gardner Bowers and Margaretta Keller Bowers, "Sterilization of the Female by Cauterization of the Uterine Cornu," *Journal of the Tennessee State Medical Association* 31, no. 10 (October 1938): 386, AVS box 14 folder 116.

61. Terry, "Anxious Slippages between 'Us' and 'Them,'" 141; Reed, *From Private Vice to Public Virtue,* 156; Terry, "Anxious Slippages between 'Us' and 'Them,'" 143.

62. Reed, *From Private Vice to Public Virtue,* 191; Rachel Maines, *The Technology of Orgasm: "Hysteria," the Vibrator, and Women's Sexual Satisfaction* (Baltimore: Johns Hopkins University Press, 1999); see also p. 181 n 59. I am indebted to Jennifer Terry for her insights into Dickinson, via e-mail correspondence.

63. Reed, *From Private Vice to Public Virtue,* 165, 161.

64. "Eugenical Sterilization at the Meeting of the AMA," *Eugenical News* 13 (August 1928): 115.

65. McCann, *Birth Control Politics,* 94.

66. Reed, *From Private Vice to Public Virtue,* 45. Dickinson himself, who became a leading medical advocate of contraception, initially refused in the 1910s to publicly join the cause because of his wife's insistence that it would damage his career; Reed, *From Private Vice to Public Virtue,* 46.

67. Mariann Norton to Frank Reid, 10 October 1941, Gosney Papers, 9.3; "Eugenic Sterilization Laws Inadequate, Study Reveals," *AMA News,* June 1960, AVS box 14, Butler file.

68. E. S. Gosney, "Annual Report of the HBF for the Year Ending Feb. 14, 1939," Gosney Papers, 1.1.

69. David Slight to HBF, 22 September 1941, Gosney Papers, 7.12.

70. Gosney to Lewis Terman, n.d., Terman Papers, SC 38, 2–11.

71. Jacob Landes, "Radio Broadcast," WKBQ, 15 April 1930, Gosney Papers, 18.2.

72. Dr. Signed Dahlstron to Gosney, n.d., Gosney Papers, 18.2.

73. Havelock Ellis to Gosney, 14 February 1930, Gosney Papers, 18.2.

74. Glen Seaman to HBF, October 1941, Gosney Papers, 8.5.

75. Roy Neal to HBF, March 1940, Gosney Papers, 8.12.

76. Arthur Bennet, M.D., to Gosney, 24 December 1929, Gosney Papers, 18.2.

77. Ellsworth Huntington to Gosney, 7 October 1940, Gosney Papers, 6.20.

78. Major Leonard Darwin to Gosney, 30 September 1929, Gosney Papers, 18.2.

79. Department of Institutions, *Biennial Report*, 1922–32; Department of Institutions, *Statistical Report*, 1934–45.

80. McCann, *Birth Control Politics*, 45.

81. Popenoe, "Success on Parole after Sterilization," 9, 2.

82. For analysis of the construction of the female "sex delinquent" in the early twentieth century, see chapter 2.

83. Popenoe, "Success on Parole after Sterilization," 16.

84. Ibid., 13.

85. E. A. Whitney and Mary McD. Shick, M.D., "Some Results of Selective Sterilization," *Journal of Psycho-Asthenics* 36 (1931): 333.

86. Butler, "Sterilization Procedure and Its Success in California Institutions," 8; F. O. Butler, "Some Results of Selective Sterilization: Discussion," *Journal of Psycho-Asthenics* 36 (1931): 336.

87. Paul Popenoe, "Eugenic Sterilization in California: Effect of Salpingectomy on the Sexual Life," in *Collected Papers on Eugenic Sterilization in California: A Critical Study of Results in 6000 Cases* (Pasadena, Calif.: Human Betterment Foundation, 1930), 1.

88. Popenoe, "Success on Parole after Sterilization," 11.

89. Paul Popenoe, "The Opinions of Some California Physicians," n.d., Gosney Papers, 12.8.

90. Ibid.

91. Questionnaire for social workers, n.d. [1926], Gosney Papers, 16.5.

92. Questionnaire for physicians, n.d. [1926], Gosney Papers, 12.8.

93. Ibid.; see chapter 2.

94. Morantz-Sanchez, *Conduct Unbecoming a Woman*, 129.

95. Questionnaire for social workers, n.d. [1926], Gosney Papers, 16.5.

96. Ibid.

97. Joel Braslow, *Mental Ills and Bodily Cures: Psychiatric Treatment in the First Half of the Twentieth Century* (Berkeley: University of California Press, 1997), 57. As I argue in chapter 4, eugenicists target the "misfit mother" as an important candidate for eugenic sterilization beginning in the 1930s.

98. Edwin Wayte to Gosney, 21 September 1926, Gosney Papers, 12.8.

99. Questionnaire for physicians, n.d. [1926], Gosney Papers, 12.8.

100. Butler claimed that "often a boy or girl would come to ask me if it wasn't time for his or her operation." F. O. Butler, "The Prevention of Mental Deficiency by Sterilization, 1949," unpublished manuscript, AVS box 14 folder 113. See also Dowbiggin, *Keeping America Sane*, 123, and Johanna Schoen, "'A Great Thing for Poor Folks': Birth Control, Sterilization, and Abortion in Public Health and Welfare in the Twentieth Century" (Ph.D. diss., University of

North Carolina, Chapel Hill, 1995). She found that in North Carolina sterilization "was not only a threat to women's reproductive autonomy but also a form of birth control very much desired by women" (101).

101. Sterilization patient #398, Gosney Papers, 17.8.

102. Sterilization patient #403, Gosney Papers, 17.8.

103. Sterilization patients #391, #359, Gosney Papers, 17.8.

104. "It Isn't Right for a Strong Man," *Birth Control Review* 10, no. 2 (February 1926): 53; "I Could Not Do It," *Birth Control Review* 10, no. 2 (February 1926): 53.

105. Chesler, *Woman of Valor,* 224.

106. Morantz-Sanchez, *Conduct Unbecoming a Woman,* 106, 113, 146; see also Nancy Theriot, *Mothers and Daughters in Nineteenth-Century America: The Biosocial Construction of Femininity* (Lexington: University Press of Kentucky, 1996), 99–100.

107. Schoen, "'A Great Thing for Poor Folks,'" iii, 101; Grossman, *Reforming Sex,* 71; Laura Briggs, "Discourses of 'Forced Sterilization' in Puerto Rico: The Problem with the Speaking Subaltern," *Differences* 10, no. 2 (1998): 46–47.

108. Sterilization patient #2053, Gosney Papers, 17.8.

109. Sterilization patient #403, Gosney Papers, 17.8.

110. Department of Institutions, *Biennial Report* (1926), 92.

111. Anonymous to Human Betterment Foundation, 8 April 1940; response [n.d.], Gosney Papers, 5.17.

112. Rosemarie Thomson, *Extraordinary Bodies: Figuring Physical Disability in American Culture and Literature* (New York: Columbia University Press, 1997), 8.

113. Stanford Vieth to HBF, n.d., Gosney Papers, 6.10. The HBF responded by confirming his anxiety. "On the surface of it," they wrote, "your case seems to present a sound reason for the operation." HBF to Stanford Vieth, 17 March [year unknown], Gosney Papers, 6.10.

114. Miss M. Walsh to HBF, n.d., Gosney Papers, 6.12.

115. Butler, "Some Results of Selective Sterilization: Discussion," 337.

116. G. B. Arnold, "A Brief Review of the First Thousand Patients Eugenically Sterilized at the State Colony for Epileptics and Feebleminded," *Journal of Psycho-Asthenics* 43 (1938): 57.

117. Elwan Adams to Gosney, 5 October 1929, Gosney Papers, 18.2.

118. Louana Siler to HBF, 7 June 1941, Gosney Papers, 6.9.

119. Joe Patterson Smith (Department of History, Illinois College) to HBF, 16 October 1940, Gosney Papers, 7.12.

120. Mary Lipton to HBF, 5 July 1943, Gosney Papers, 7.12.

121. Rabbi Rudolph Coffee to Gosney, 30 August 1929, Gosney Papers, 18.2.

122. Dickinson, "Sterilization without Unsexing: Its Relative Simplicity."

Chapter Four

1. For more on the shift in American society from individualism to social responsibility, see Dawley, *Struggles for Justice,* 5.

2. William Leuchtenberg, *The Perils of Prosperity 1914–1932* (Chicago: University of Chicago Press, 1958), 249; Dixon Wecter, *The Age of the Great Depression,*

1929–1941 (New York: Macmillan, 1948), 31–32; Caroline Bird, *The Invisible Scar* (New York: McKay, 1966), 57–61.

3. Theodore Caplow, Howard Bahr, Bruce Chadwick, Reuben Hill, and Margaret Williamson, *Middletown Families: Fifty Years of Change and Continuity* (Minneapolis: University of Minnesota Press, 1982), 120; Bird, *The Invisible Scar,* 61, 58.

4. Wecter, *The Age of the Great Depression,* 31; Caplow et al., *Middletown Families,* 120; see also Winona Morgan, *The Family Meets the Depression: A Study of a Group of Highly Selected Families* (Westport, Conn.: Greenwood Press, 1939), 55–68.

5. Dawley, *Struggles for Justice,* 383; McCann, *Birth Control Politics,* 187.

6. Mintz and Kellogg, *Domestic Revolutions,* 134, 136–37.

7. Leuchtenberg, *The Perils of Prosperity,* 249; Bird, *The Invisible Scar,* 283; Studs Terkel, *Hard Times: An Oral History of the Great Depression* (New York: Pantheon Books, 1970), 389.

8. Bird, *The Invisible Scar,* 51; Susan Ware, *Holding Their Own: American Women in the 1930s* (Boston: Twayne, 1982), 7; Bird, *The Invisible Scar,* 53.

9. Bird, *The Invisible Scar,* 52; Fred Hogue, "Social Eugenics," *Los Angeles Times Sunday Magazine,* 29 November 1936, p. 31; Louis Dublin, "The Excesses of Birth Control," in *Problems of Overpopulation,* ed. Margaret Sanger (New York, 1926), 186, quoted in Reed, *From Private Vice to Public Virtue,* 208.

10. Dr. Alfred Mjoen quoted in "Scientist Says Modern Women Ruining Race," *Watsonville Pajaronian,* 3 March 1930, p. 1.

11. Cravens, *The Triumph of Evolution,* 222–23.

12. Degler, *In Search of Human Nature,* 141.

13. Haller, *Eugenics,* 179. For historians who argue that eugenics was in serious decline by the 1930s, see the Introduction, n. 11.

14. Dawley, *Struggles for Justice,* 5

15. "Even Eugenists Carry Germ of Feeblemindedness and Sterilization Is Folly, Asserts N.Y.U. Professor," *New York World Telegraph,* 29 April 1934; see also Kevles, *In the Name of Eugenics,* 165.

16. Degler, *In Search of Human Nature,* 150–51.

17. Larson, *Sex, Race, and Science,* 120; Grether, "Sterilization and Eugenics," 184. Dowbiggin notes that eugenicists in the interwar period "often emphasized environment as well as heredity," but he does not use this observation to challenge the existing historiography regarding the decline of eugenics; Dowbiggin, *Keeping America Sane,* 124. For more specific sterilization data, see Philip Reilly, *The Surgical Solution: A History of Involuntary Sterilization in the United States* (Baltimore: Johns Hopkins University Press, 1991), ch. 7.

18. Butler, "Some Results of Selective Sterilization: Discussion," 336.

19. Hoover quoted in James Hay Jr., "'A Wise Son Maketh a Glad Father': The President's November Conference on Child Health Will Work to Develop Our Greatest Asset," *World's Work* 59 (October 1930): 73. Birthright began under the leadership of Mariann Olden as the New Jersey Sterilization League in 1937. The name was changed to Birthright in 1943. Butler left his position as medical superintendent of the Sonoma State Home to become medical director for Birthright in 1949. In that same year, the name was changed to the Association for Voluntary Sterilization (AVS), which remains its name today. After Gosney's death and the end of the HBF in 1942, the AVS became the sole organization

conducting surveys of institutional sterilization in the United States. See Reilly, *The Surgical Solution*, 135–37; William VanEssendelft, "A History of the Association for Voluntary Sterilization: 1935–1964" (Ph.D. diss., University of Minnesota, 1978).

20. Dawley, *Struggles for Justice*, 309.

21. Ray Lyman Wilbur, "Toward a Better Child Life," *Review of Reviews* 83 (January 1931): 55.

22. Alan Brinkley, *The Unfinished Nation: A Concise History of the American People* (New York: McGraw-Hill, 1993), 637. For more on the impact of radio on popular culture in the 1920s and 1930s, see Lizabeth Cohen, *Making a New Deal: Industrial Workers in Chicago, 1919–1939* (New York: Cambridge University Press, 1990). She finds that approximately 80 percent of middle-class and 50 percent of working-class Chicagoans owned radios in 1930 (134).

23. "President Hoover's Address on Nation's Children," *New York Times*, 20 November 1930, p. 2.

24. Haven Emerson, M.D., "From Promoters to Parents: What Can We Expect from the White House Conference?" *Survey* 67 (November 1931): 191–92.

25. Dawley, *Struggles for Justice*, 308.

26. E. R. Johnstone, "Report of the Committee on Mental Deficiency of the White House Conference," *Journal of Psycho-Asthenics* 36 (1931): 341.

27. Ibid., 347.

28. "Annual Report of the Human Betterment Foundation" 9 February 1932, Terman Papers, SC 38, 2–13.

29. Quoted in Hay, "'A Wise Son Maketh a Glad Father,'" 73.

30. Stepan, *The Hour of Eugenics*, 195. Reilly notes that because of the destruction of Nazi war records, we will never know how many people were sterilized; estimates have gone as high as 3,500,000. Reilly, *The Surgical Solution*, 109.

31. Eugen Fischer to Gosney, 18 February 1930, trans. Stefan Paula, Gosney Papers, 18.2.

32. Hermann Simon to Gosney, 20 April 1930, Gosney Papers, 18.2.

33. Dr. Marie Kopp, "Draft—as Read before Eugenic Association," 7 May 1936, AVS box 13 folder 111. For more on the links between eugenics in the United States and Germany, see Kühl, *The Nazi Connection*.

34. Paul Popenoe, "The German Sterilization Law," *Journal of Heredity* 25, no. 7 (July 1934): 257.

35. Ibid., 260, 257.

36. Fred Hogue, "Social Eugenics," *Los Angeles Times Sunday Magazine*, 9 February 1936, p. 31.

37. Gosney to Frank Reid, 9 September 1940, Gosney Papers, 1.2. Emphasis in original.

38. Paul Popenoe to Mariann Olden, 5 May 1945, AVS box 2 folder 12.

39. Dawley, *Struggles for Justice*, 310–16; Cott, *The Grounding of Modern Feminism*, 146–47; Cohen, *Making a New Deal*.

40. Cott, *The Grounding of Modern Feminism*, 147.

41. For more on the increasing significance of the "normal" in twentieth-century American society, see Baynton, *Forbidden Signs*.

42. See Joan Jacobs Brumberg, *The Body Project: An Intimate History of American Girls* (New York: Random House, 1997); Roland Marchand, *Advertising the*

American Dream: Making Way for Modernity, 1920–1940 (Berkeley: University of California Press, 1985).

43. Frederick Osborn, "Implications of the New Studies in Population and Psychology for the Development of Eugenic Philosophy," *Eugenical News* 22, no. 6 (1937): 107, quoted in Mehler, "A History of the American Eugenics Society," 124.

44. Edgar Doll, "Current Thoughts on Mental Deficiency," *Journal of Psycho-Asthenics* 41 (June 1936): 44.

45. Gladys Schwesinger, "Sterilization and the Child," 1–2, extracts from a paper read at the Sixty-Third Annual Meeting of the New Jersey Health and Sanitary Association, 10 December 1937, reprinted from *"Health Progress," Official Publication of the New Jersey Health and Sanitary Association, Inc.,* Gosney Papers, 11.7.

46. Mintz and Kellogg, *Domestic Revolutions,* 144–49.

47. "Race Degeneration Seen for America," *New York Times,* 21 June 1932, AVS box 15 folder 123.

48. Reilly, *The Surgical Solution,* 97.

49. Larson, *Sex, Race, and Science,* 119.

50. Paul Popenoe and Norman Fenton, "Sterilization as a Social Measure," *Journal of Psycho-Asthenics* 41 (1936): 60–65.

51. Morantz-Sanchez, *Conduct Unbecoming a Woman,* 195.

52. *Literary Digest,*19 December 1936, p. 13.

53. Kevles, *In the Name of Eugenics,* 112.

54. Herbert Ray, "The Law and the Surgeon's Liability in Human Sterilization Operations," 8 June 1935, 7, AVS box 24, HBF file.

55. "Three Warrants Issued in Ann Hewitt Case," *New York Times,* 5 February 1936, p 3.

56. I am interested here in how the Cooper Hewitt trial was depicted in the press, and I am suggesting that the construction of the narratives themselves resonated in the 1930s with readers concerned about the changing roles of mothers and daughters in American society. Judith Walkowitz, in *City of Dreadful Delight: Narratives of Sexual Danger in Late-Victorian London* (Chicago: University of Chicago Press, 1992), illuminates the meaning of sexual danger in Victorian London by analyzing W. T. Stead's "Maiden Tribute of Modern Babylon" in the *Pall Mall Gazette.* She explains, "By focusing on narrative, I hope to explore how cultural meanings around sexual danger were produced and disseminated in Victorian society, and what were their cultural and political effects. Narratives of the 'real,' such as history and news reporting, impose a formal coherence on events: they 'narrativize' data into a coherent well-made tale, converting chaotic experience into meaningful moral drama" (83). I am suggesting that for America in the 1930s the Cooper Hewitt trial was such a "meaningful moral drama," which is why it received widespread coverage.

57. "Charges She Was Sterilized," *New York Sun,* 6 January 1936, AVS box 14 folder 120.

58. Ibid.

59. Ibid.

60. Gordon, *Heroes,* 151–52.

61. "Police to Study Ann Hewitt's Operation Case," *New York Tribune,* 8 January 1936, AVS box 14 folder 120.

62. "Hewitt Heiress Says Mother Had Her Sterilized to Get Estate," *New York Tribune,* 7 January 1936, AVS box 14 folder 120.

63. "Calls Ann Hewitt Overly Romantic," *New York Times,* 10 January 1936, p. 2.

64. "Accuses Mother in Court," *New York Times,* 24 January 1936, p. 40.

65. "Mother Cites 'Erotic Letters' of Ann Hewitt," *New York Tribune,* 10 January 1936, AVS box 14 folder 120. Emphasis added.

66. A copy of the Scally report was enclosed in a letter to Popenoe from I. M. Golden, the San Francisco attorney who defended the surgeons in the August trial, dated 23 May 1936, Gosney Papers, 11.9.

67. "Alienists Line Up in Hewitt Contest," *New York Times,* 9 January 1936, 2.

68. "Judge Postpones Decision on Sterilization Charges," *Pasadena Post,* 24 January 1936, Gosney Papers, 11.9.

69. "Psychiatry Claims Another Victim," *Sacramento Bee,* 27 January 1936, Gosney Papers, 11.9.

70. "Judge Postpones Decision on Sterilization Charges," *Pasadena Post,* 24 January 1936, Gosney Papers, 11.9.

71. "Hewitt Case Judge Told His Mental Age Is 12," *New York Tribune,* 19 February 1936, AVS box 14 folder 120.

72. "Judge Given Child Rating," *Los Angeles Times,* 19 February 1936, Gosney Papers, 11.9.

73. "Two Doctors Held in Ann Hewitt Case," *New York Times,* 20 February 1936, p. 3.

74. Kevles, *In the Name of Eugenics,* 345–46.

75. "Doctor Upholds Mental Fitness of Miss Hewitt," *Herald Tribune,* 9 January 1936, AVS box 14 folder 120.

76. "$500,000 Operation," *Time,* 20 January 1936, 42–45.

77. "Stepbrother Denies Cruelty," *New York Times,* 11 January 1936.

78. "Tell of Hewitt Operation," *New York Sun,* 9 January 1936, AVS box 14 folder 120.

79. Ibid.

80. "Charges She Was Sterilized," *New York Sun,* 6 January 1936, AVS box 14 folder 120.

81. "Calls Ann Hewitt Overly Romantic," *New York Times,* 10 January 1936; I. M. Golden to Paul Popenoe, 23 May 1936, Gosney Papers, 11.9.

82. "Tell of Hewitt Operation," *New York Sun,* 9 January 1936, AVS box 14 folder 120.

83. Popenoe to Golden, 26 May 1936, Gosney Papers, 11.9.

84. See Fass, *The Damned and the Beautiful;* Odem, *Delinquent Daughters;* Lunbeck, *The Psychiatric Persuasion;* Kathy Peiss, *Cheap Amusements: Working Women and Leisure in Turn-of-the-Century New York* (Philadelphia: Temple University Press, 1986); Kunzel, *Fallen Women, Problem Girls;* Alexander, *The "Girl Problem";* Meyerowitz, *Women Adrift.*

85. E. S. Gosney, quoted in Fred Hogue, "Social Eugenics," *Los Angeles Times Sunday Magazine,* 1 March 1936, p. 30.

86. D'Emilio and Freedman, *Intimate Matters,* 255; Helen Rodriguez-Trìas, "Sterilization Abuse," in *Biological Woman—the Convenient Myth: A Collection of Feminist Essays and a Comprehensive Bibliography* (Cambridge, Mass.: Schenkman Books, 1982). According to Davis, 24 percent of all Native American women of

childbearing age and 35 percent of Puerto Rican women of childbearing age had been sterilized by 1976. Angela Davis, *Women, Race, and Class* (New York: Vintage Books, 1983), 218–19. More recent scholarship, however, suggests the need to reevaluate the portrayal of sterilized women as merely victims of social control. See chapter 2; and Briggs, "Discourses of 'Forced Sterilization' in Puerto Rico."

87. "Ann Cooper Hewitt Sues Her Mother," *New York Times,* 7 January 1936, p. 3.

88. "Dr. Boyd's Card," copy of surgeon's record enclosed in letter to Paul Popenoe from I. M. Golden, 23 May 1936, Gosney Papers, 11.9.

89. "$500,000 Operation," *Time,* 20 January 1936, 45.

90. "Ann Cooper Hewitt Sues Her Mother," *New York Times,* 7 January 1936, p. 3. The issue of parental rights in acting in the "best interest of the child" is still debated today in California. On February 21, 1998, the *San Jose Mercury News* reported that a couple had paid two strangers to take their disobedient son out of the house in the middle of the night and ship him off for a year of disciplinary action at a Jamaican reform school. An Alameda County Superior Court judge ruled that they were acting within their parental rights, a ruling "consistent with decades of law that gives parents broad latitude in rearing their children as they see fit."

91. I. M. Golden to Paul Popenoe, 23 May 1936, Gosney Papers, 11.9.

92. Popenoe to Golden, 26 May 1936, Gosney Papers, 11.9.

93. Mehler, "A History of the American Eugenics Society," 260.

94. Ellsworth Huntington, *Tomorrow's Children: The Goal of Eugenics* (New York: Wiley, 1935), 7.

95. Ibid., 63.

96. Golden to Popenoe, 28 May 1936, Gosney Papers, 11.9.

97. Justin Miller and Gordon Dean, "Liability of Physicians for Sterilization Operations," *Journal of the American Bar Association,* March 1930, 158–61, Gosney Papers, 19.6.

98. Hartley F. Peart, Esq., "Vasectomy and Salpingectomy under California Law," *California and Western Medicine,* May-June 1941, 6, AVS box 73, California Department of Mental Hygiene folder.

99. "Heiress Accuses Two San Francisco Doctors," *San Francisco Examiner,* 14 August 1936, Gosney Papers, 11.9.

100. "Hewitt Trial Plea Mapped," *San Francisco News,* 20 August 1936, Gosney Papers, 11.9.

101. Butler to Gosney, 30 January 1939, Gosney Papers, 11.9.

102. Popenoe to Golden, 26 May 1936, Gosney Papers, 11.9.

103. Popenoe to Golden, 31 August 1936, Gosney Papers, 11.9.

104. *Literary Digest,* 19 December 1936, p. 13.

105. Reilly, *The Surgical Solution,* 125.

106. Fred Hogue, "Social Eugenics," *Los Angeles Times Sunday Magazine,* 13 September 1936, p. 31.

107. Fred Hogue, "Social Eugenics," *Los Angeles Times Sunday Magazine,* 6 December 1936, p. 31.

108. Fred Hogue, "Social Eugenics," *Los Angeles Times Sunday Magazine,* 13 December 1936, p. 29.

109. Quoted in Peart, "Vasectomy and Salpingectomy under California Law," 7.

110. Roswell Johnson, "Comment on Dr. F. O. Butler's Paper at Railway Surgeon's Meeting, Los Angeles," 7 October 1938, Gosney Papers, 11.7.

Chapter Five

1. Elaine May, *Great Expectations: Marriage and Divorce in Post-Victorian America* (Chicago: University of Chicago Press, 1980), 2; Mintz and Kellogg, *Domestic Revolutions,* 171.

2. Mintz and Kellogg, *Domestic Revolutions,* 136; Karen Anderson, *Wartime Women: Sex Roles, Family Relations, and the Status of Women during World War II* (Westport, Conn.: Greenwood Press, 1981), 106–11; Gordon, *Woman's Body, Woman's Right,* 352.

3. Alice Kessler-Harris, *Out to Work: A History of Wage-Earning Women in the United States* (New York: Oxford University Press, 1982), 259.

4. Mintz and Kellogg, *Domestic Revolutions,* 161.

5. Eugenic interest in sterilizing the "unfit" continued, but many eugenicists believed they could exercise more influence by focusing on positive eugenics. Positive eugenics gained popularity in the 1920s as some eugenicists initiated "Fitter Families for Future Firesides" contests at midwestern state fairs (see Rydell, *World of Fairs*). However, interest in marriage and family counseling as a way to promote procreation did not emerge until the 1930s.

6. Wini Breines, *Young, White, and Miserable: Growing Up Female in the Fifties* (Boston: Beacon Press, 1992), 50.

7. Mintz and Kellogg, *Domestic Revolutions,* 179, 178; Susan Householder Van Horn, *Women, Work, and Fertility* (New York: New York University Press, 1988), 92; Elaine May, *Homeward Bound: American Families in the Cold War Era* (New York: Basic Books, 1988); Robert Griswold, *Fatherhood in America: A History* (New York: Basic Books, 1993), 188–89; Breines, *Young, White, and Miserable,* 49.

8. In *Barren in the Promised Land,* Elaine May, noted historian of families in the postwar era, argues that positive eugenics "never took hold among the population at large." Referring to the turn-of-the-century race-suicide panic initiated by Teddy Roosevelt, she writes that "if the American people did not respond to pronatalist moral suasion, little could be done to boost the birthrate of the 'best' stock; . . . the crusade for positive eugenics failed" (92). Her restricted definition of eugenics as a hereditary science does not allow her to see its later influence on postwar pronatalism.

9. Fred Hogue, "Social Eugenics," *Los Angeles Times Sunday Magazine,* 25 October 1936, p. 31.

10. Havelock Ellis, "The History of Marriage," in *Eonism and Other Supplementary Studies* (1928), quoted in Robinson, *The Modernization of Sex,* 34.

11. Dr. Reed to C. M. Goethe, quoted in letter from Goethe to Terman, 4 May 1950, Terman Papers, SC 38, 4–12.

12. D'Emilio and Freedman, *Intimate Matters,* 229; Ellen Kay Trimberger, "Feminism, Men, and Modern Love: Greenwich Village, 1900–1925," in *Powers of Desire: The Politics of Sexuality,* ed. Ann Snitow, Christine Stansell, and Sharon Thompson (New York: Monthly Review Press, 1983), 131–52.

13. Floyd Dell, *Janet March* (New York: Knopf, 1923), 202.

14. D'Emilio and Freedman, *Intimate Matters*, 256–57.

15. Dell, *Janet March*, 209.

16. Ibid., 204, 210, 207.

17. Ibid., 457.

18. Review of *Janet March*, by Floyd Dell, *Bookman* 58 (December 1923): 453.

19. Florence Guy Seabury, "Stereotypes," in *Our Changing Morality: A Symposium*, ed. Freda Kirchway (New York: Boni, 1924), 225.

20. Dell, *Janet March*, 192.

21. Benjamin Barr Lindsey and Wainwright Evans, *Companionate Marriage* (New York: Boni & Liveright, 1927); Christina Simmons, "Companionate Marriage and the Lesbian Threat," in *Women and Power in American History: A Reader*, vol. 2, ed. Kathryn Kish Sklar and Thomas Dublin (Englewood Cliffs, N.J.: Prentice-Hall, 1991), 185; D'Emilio and Freedman, *Intimate Matters*, 266.

22. Simmons, "Companionate Marriage," 192, Cott, *The Grounding of Modern Feminism*, 174.

23. Olive Higgins Prouty, *Stella Dallas* (New York: Grosset & Dunlap, 1923), 85, 87.

24. Ibid., 92, 100–101.

25. Ibid.,150–51.

26. Ibid., 249.

27. Ellsworth Huntington, *Tomorrow's Children: The Goal of Eugenics* (New York: Wiley, 1935), 60.

28. Harry Laughlin, "Further Studies on the Historical and Legal Development of Eugenical Sterilization in the United States," *Journal of Psycho-Asthenics* 41 (1936): 96.

29. "Regular Meeting," 4 November 1929, AES Southern California records, Popenoe file, Davenport Papers, #1.

30. "Regular Meeting," 13 October 1931, 11 April 1932, 9 May 1932, AES Southern California records, Popenoe file, Davenport Papers, #1.

31. "Regular Meeting," 4 November 1929, Popenoe file, Davenport Papers, #1.

32. Paul Popenoe, "How Can Colleges Prepare Their Students for Marriage and Parenthood?" *Journal of Home Economics* 22, no. 3 (1930): 169, 170, 175, 178.

33. "Summary of the Proceedings of the Conference on Recreation and the Use of Leisure Time in Relation to Family Life," 22 January 1937, Conferences file, AES Papers, American Philosophical Society, Philadelphia.

34. "Co-Recreation or Recreation for Young Men and Women Together" (paper presented at the Conference on Recreation, AES, 22 January 1937), Conferences file, AES Papers; AES, "Summary of the Proceedings of the Conference on Recreation and the Use of Leisure Time," 22 January 1937, Conferences file, AES Papers.

35. "Co-Recreation or Recreation for Young Men and Women Together."

36. "Birth Control's Opportunity to Strengthen Our Human Resources—Our Population," 6 July 1940, Planned Parenthood Federation of America (PPFA) Papers, Sophia Smith Collection, Smith College Library, quoted in Gordon, *Woman's Body, Woman's Right*, 352; Dr. Richard N. Pierson, speech at the 1941 annual meeting of the Birth Control Federation, PPFA Papers, quoted in Gordon, *Woman's Body, Woman's Right*, 340; Henry Pratt Fairchild, speech at the 1940 an-

nual meeting of the Birth Control Federation, PPFA Papers, quoted in Gordon, *Woman's Body, Woman's Right,* 285.

37. McCann, *Birth Control Politics,* 80; Reed, *From Private Vice to Public Virtue,* 148; Margaret Marsh and Wanda Rommer, *The Empty Cradle: Infertility in America from Colonial Times to the Present* (Baltimore: Johns Hopkins University Press, 1996), 149; David Loth, "The Most Unforgettable Character I've Met," *Reader's Digest,* April 1952, AVS box 15, Dickinson file; Robert Dickinson, "Premarital Examination as Routine Preventive Gynecology," paper read before the American Gynecological Society, Fifty-Third Annual Meeting, Washington, D.C., 30 April–2 May 1928, p. 2, AVS box 15, Dickinson file.

38. Dickinson, "Premarital Examination," 2, 5; Louise Stevens Bryant to George Kosmak and Howard Taylor Jr., 20 Nov 1934, AVS box 13 folder 106.

39. Robert Dickinson, *A Thousand Marriages: A Medical Study of Sex Adjustment* (Baltimore: Williams & Wilkins, 1931). Quotes are from Robert Dickinson, "Medical Analysis of a Thousand Marriages," paper read before the Section on Obstetrics, Gynecology and Abdominal Surgery at the Eighty-Second Annual Session of the American Medical Association, Philadelphia, 10 June 1931, pp. 3, 5, AVS box 15, Dickinson file.

40. Smith-Rosenberg, "Discourses of Sexuality and Subjectivity," 270–71.

41. Chauncey, "From Sexual Inversion to Homosexuality," 90; Smith-Rosenberg, "Discourses of Sexuality and Subjectivity," 270–71.

42. Terry, "Anxious Slippages between 'Us' and 'Them,'" 138, 139.

43. Ibid., 148.

44. "News Bulletin: Manual of Information and Direction for Use of M-F Test," McGraw-Hill Books, n.d., Terman Papers, SC 38, 15–32.

45. Terman and Miles, *Sex and Personality,* 5.

46. Ibid., 482–554.

47. "Terman & Miles: Sex and Personality," book review in *Clinical Medicine and Surgery,* October 1937, 477–78, Terman Papers, SC 38, 15–32.

48. Havelock Ellis, "How Do Men and Women Differ?" *New Chronicle,* 8 April 1937, Women's Page, Terman Papers, SC 38, 15–32.

49. Terry, "Anxious Slippages between 'Us' and 'Them,'" 139.

50. Dr. Rosanoff to Lewis Terman 5 March 1929, Terman Papers, SC 38, 12–31.

51. Lewis Terman to Dr. Rosanoff, 14 March 1929, Terman Papers, SC 38, 12–31.

52. As Terry points out, the Sex Variant study made the same problematic assumptions about homosexuality and gender acquisition; Terry, "Anxious Slippages between 'Us' and 'Them,'" 148–49.

53. Lewis Terman to Dr. J. Harold Williams, 21 January 1929, Terman Papers, SC 38, 12–31.

54. Terman and Miles, *Sex and Personality,* 248.

55. Ibid., 345–46.

56. D'Emilio and Freedman, *Intimate Matters,* 294.

57. Terman and Miles, *Sex and Personality,* 467–68.

58. William E. Allfather to Lewis Terman, 16 January 1939, Terman Papers, SC 38, 15–32.

59. Lewis Terman and E. Lowell Kelly, "Case Studies of Homosexual Males," in Terman and Miles, *Sex and Personality,* 320.

60. William E. Allfather to Lewis Terman, 16 January 1939, Terman Papers, SC 38, 15–32; Terman to Allfather, 23 January 1939, Terman Papers, SC 38, 15–32. Another young man, who considered Terman "an authority on the subject of Homosexuality" after reading *Sex and Personality,* wrote for advice on how he could "change his condition." He had been given a series of injections by Dr. Sydney Smith in Oakland, California, but they didn't help; he then began a prescription of "Antuitrin S," which also did not work. He therefore asked Terman for more suggestions. Terman, however, did not "feel competent" to advise him in regard to physical treatment. Kenneth G. Vincent to Lewis Terman, 8 October 1937; response, 13 October 1937, Terman Papers, SC 38, 15–33.

61. Michael Kimmel, *Manhood in America: A Cultural History* (New York: Free Press, 1996), 209; Cott, *The Grounding of Modern Feminism,* 154.

62. Breines, *Young, White, and Miserable,* 30–34.

63. Popenoe, "How Can Colleges Prepare Their Students for Marriage and Parenthood?" 172.

64. Minton, *Lewis M. Terman,* 176.

65. Lewis Terman, *Psychological Factors in Marital Happiness* (New York: McGraw-Hill, 1938), 146.

66. Minton, *Lewis M. Terman,* 183.

67. Henry C. Link, "The Way to Marital Happiness as Diagrammed by Lewis Terman—a Review," *Journal of Heredity,* n.d., 286–87, Terman Papers, SC 38, 4–1.

68. J. I. Cohen, "Psychological Factors," *Eugenics Review* (April 1939), Terman Papers, SC 38, 4–1.

69. Robert Dickinson to Lewis Terman, 2 December 1938, Terman Papers, SC 38, 4–1.

70. Havelock Ellis to Lewis Terman, 11 November 1938, Terman Papers, SC 38, 4–1.

71. Lewis Terman to Paul Popenoe, 1 November 1934, Terman Papers, SC 38, 4–1.

72. Betty Hannah Hoffman, "The Man Who Saves Marriages," *Ladies' Home Journal,* September 1960, 125.

73. "The Department of Education, Institute of Family Relations, All-Day Regional Conference on Family Relations," program pamphlet, 2 December 1933, Terman Papers, SC 38, 2–7.

74. Paul Popenoe, "The Writings of Dr. Popenoe," *Family Life: Special Memorial Issue* 39, no. 5 (September and October 1979): 9; Hoffman, "The Man Who Saves Marriages," 71.

75. Grossman, *Reforming Sex,* 9–10.

76. Marie Kopp, "The Development of Marriage Consultation Centers as a New Field of Social Medicine," reprinted from the *American Journal of Obstetrics and Gynecology* 26, no. 1 (July 1933), 7, AVS box 13 folder 111; Grossman, *Reforming Sex,* 141.

77. Chesler, *Woman of Valor,* 304; Gordon, *Woman's Body, Woman's Right,* 365; "Memorandum of visit of E. S. Gosney to New York City, May 3–6 1935," Gosney Papers, 18.2; Gordon, *Woman's Body, Woman's Right,* 359.

78. "Dr. Fishbein Tells Health Progress," *L.A. Herald and Express,* 12 April 1935, Gosney Papers, 13.4; Robert Laidlaw, "Introduction," in *Successful Marriage: A*

Modern Guide to Love, Sex, and Family Life, ed. Morris Fishbein (Garden City, N.Y.: Doubleday, 1963), xiv.

79. May, *Great Expectations,* 75.

80. Title of *Ladies' Home Journal* article, September 1960.

81. Quoted in James S. Hirsch, "Happy Endings: This Marriage, Too, Is Saved by Therapy," *Wall Street Journal,* 14 January 1992, p. A1.

82. Hoffman, "The Man Who Saves Marriages," 123.

83. Paul Popenoe, "What Science Can Do for Matrimony" (address delivered before the Southern California division of the American Eugenics Society, Los Angeles Public Library, 18 April 1940), Gosney Papers, 19.2.

84. Popenoe, "The Writings of Dr. Popenoe," 11.

85. Popenoe, "What Science Can Do for Matrimony."

86. "The Institute of Family Relations," reprint from April 1931 *Journal of Juvenile Research,* Gosney Papers, 19.2.

87. "The Department of Education, Institute of Family Relations, All-Day Regional Conference on Family Relations."

88. American Institute of Family Relations, "Premarital Conference," [n.d.], Popenoe file, Davenport Papers, #4.

89. Paul Popenoe, "Some Biological Differences between Men and Women," pamphlet used in Graduate Workshop in Techniques of Marriage and Family Counseling, American Institute of Family Relations, August 1974, National Council on Family Relations Records, box 64, American Institute of Family Relations (AIFR) file, Social Welfare History Archives, University of Minnesota, Minneapolis.

90. Popenoe, "How Can Colleges Prepare Their Students for Marriage and Parenthood?" 169.

91. Paul Popenoe, "Are Homosexuals Necessary?" publication 542, American Institute of Family Relations, [n.d.], AVS box 29, AIFR file.

92. Peter and Barbara Wyden, quoted in Paul Popenoe, "Family Strength and Mental Health," publication 540, American Institute of Family Relations, [n.d.], AVS box 29, AIFR file.

93. Popenoe, "What Science Can Do for Matrimony."

94. Jennifer Scanlon, *Inarticulate Longings: The Ladies' Home Journal, Gender, and the Promises of Consumer Culture* (New York: Routledge, 1995), 8, 13–14.

95. Huntington, *Tomorrow's Children,* 60.

96. The *Journal's* circulation was one million in 1904 and had increased to five million by 1992. Scanlon, *Inarticulate Longings,* 4; Hirsch, "Happy Endings," A1.

97. "Can This Marriage Be Saved?" *Ladies' Home Journal,* January 1953, 40.

98. Paul Popenoe, "'I Was Going to Get a Divorce,'" *Ladies' Home Journal,* July 1941, 80.

99. Paul Popenoe, "Marriage Is for Adults Only," *Ladies' Home Journal,* February 1942, 78.

100. Paul Popenoe, "Keep Your Individuality in Marriage," *Ladies' Home Journal,* November 1942, 18.

101. Albert Deutsch, "Danger! Venereal Disease," *Nation* 161 (September 22, 1945): 285.

102. Brandt, *No Magic Bullet,* 168.

103. Anderson, *Wartime Women,* 106–11.

104. Robert B. Westbrook, "Fighting for the American Family: Private Interests and Political Obligation in World War II," in *The Power of Culture: Critical Essays in American History,* ed. Richard Fox and T. J. Jackson Lears (Chicago: University of Chicago Press, 1993); May, *Barren in the Promised Land,* 131.

105. Paul Popenoe, "Now Is the Time to Have Children," *Ladies' Home Journal,* July 1942, 60.

106. Ibid., 61; Robinson, *The Modernization of Sex,* 173.

107. Popenoe, "Now Is the Time to Have Children," 61.

108. Paul Popenoe, "Meet an Engaged Couple," *Ladies' Home Journal,* August 1943, 79.

109. Ibid.

110. Ibid.

111. "News and Notes of the AIFR," February 1942, Gosney Papers, 5.8.

112. Popenoe, "Meet an Engaged Couple," 79.

113. Ibid.

114. "Books versus Babies," *Newsweek,* 14 January 1946, 79, quoted in May, *Barren in the Promised Land,* 133.

115. Louisa Randall Church, "Parents: Architects of Peace," *American Home,* November 1946, 18–19, quoted in May, *Barren in the Promised Land,* 132.

116. Kimmel, *Manhood in America,* 245; Breines, *Young, White, and Miserable,* 52; James T. Patterson, *Grand Expectations: The United States, 1945–1974* (New York: Oxford University Press, 1996), 77, 79.

117. Patterson, *Grand Expectations,* 77, 78.

118. John Modell, *Into One's Own: From Youth to Adulthood in the United States, 1920 to 1975* (Berkeley: University of California Press, 1989), 261.

119. Rickie Solinger, *Wake Up Little Susie: Single Pregnancy and Race before Roe v. Wade* (New York: Routledge, 1992), 100.

120. Ibid.

121. Margery Rosen, "Introduction," in Margery Rosen and the Editors of the *Ladies' Home Journal, Can This Marriage Be Saved? Real-Life Cases from the Most Popular, Most Enduring Women's Magazine Feature in the World* (New York: Workman, 1994), 1; Sonya Friedman, "Foreword," in Rosen and the Editors of the *Ladies' Home Journal, Can This Marriage Be Saved?,* vii; Paul Popenoe, "Can This Marriage Be Saved?" *Ladies' Home Journal,* January 1953, 40.

122. Rosen, "Introduction," 1–2; "Can This Marriage Be Saved?" *Ladies' Home Journal,* April 1954, 122; "Can This Marriage Be Saved?" *Ladies' Home Journal,* January 1953, 40.

123. "Can This Marriage Be Saved?" *Ladies' Home Journal,* May 1954, 59, 94, 97.

124. Marsh and Rommer, *The Empty Cradle,* 196–97, 205; Breines, *Young, White, and Miserable,* 31.

125. Marsh and Rommer, *The Empty Cradle,* 186.

126. Mintz and Kellogg, *Domestic Revolutions,* 198.

127. Quoted in Breines, *Young, White, and Miserable,* 59.

128. Popenoe, "The Writings of Dr. Popenoe," 11.

129. See May, *Homeward Bound.* The 1950s also witnessed a significant shift in adoption policies. As Solinger notes in *Wake Up Little Susie,* white unwed mothers were just about forced to put their babies up for adoption, and only

black unwed mothers were encouraged to keep their children. The emphasis on a two-parent household and the stigma of unwed motherhood, so pronounced in the 1950s, were also a result of the profamily eugenic campaign.

Epilogue

1. *Parenthood,* 124 min., Universal Pictures, Universal City, Calif., 1989.

2. Breines, *Young, White, and Miserable,* 50.

3. Dickinson, "Premarital Examination," 2; David Popenoe, *Life without Father: Compelling New Evidence That Fatherhood and Marriage Are Indispensable for the Good of Children and Society* (Cambridge, Mass.: Harvard University Press, 1996), 203.

4. May, *Barren in the Promised Land,* 214; Susan Faludi, *Backlash: The Undeclared War against American Women* (New York: Crown, 1991), 27–30.

5. Ben J. Wattenberg, *The Birth Dearth* (New York: Pharos Books, 1987), quoted in May, *Barren in the Promised Land,* 215; Herrnstein quote in Faludi, *Backlash,* 32.

6. Faludi, *Backlash,* 34–35.

7. Ibid., 238, 236.

8. Stephanie Coontz, *The Way We Really Are: Coming to Terms with America's Changing Families* (New York: Basic Books, 1997), 5; Judith Stacey, *In the Name of the Family: Rethinking Family Values in the Postmodern Age* (Boston: Beacon Press, 1996), 101.

9. Barbara Dafoe Whitehead, "Dan Quayle Was Right," *The Atlantic* 271, no. 4 (April 1993): 47–84; Stacey, *In the Name of the Family,* 54.

10. Ibid., 53–54.

11. Institute for American Values, "Introduction and Overview," http://www.americanvalues.org, 10 December 2000.

12. Stacey, *In the Name of the Family,* 13; Popenoe, *Life without Father,* 2, 192; Stacey, *In the Name of the Family,* 13; Coontz, *The Way We Really Are,* 7.

13. Stacey, *In the Name of the Family,* 94.

14. Ibid., 54–55, 93–95.

15. Popenoe, *Life without Father,* 14, 24, 40, 28, 35, 63, 52.

16. Ibid., 74, 75, 174, 175.

17. Popenoe, "How Can Colleges Prepare Their Students for Marriage and Parenthood?" 170–75.

18. Popenoe, *Life without Father,* 208.

19. Institute for American Values, "Introduction and Overview"; Popenoe, *Life without Father,* 200.

20. Institute for American Values, "Introduction and Overview."

21. Popenoe, "How Can Colleges Prepare Their Students for Marriage and Parenthood?" 172; Popenoe, *Life without Father,* 139.

22. Popenoe, *Life without Father,* 210, 212.

23. Ibid., 197, 192.

Selected Bibliography

Manuscript Collections

MINNEAPOLIS: SOCIAL WELFARE HISTORY ARCHIVES, UNIVERSITY OF MINNESOTA

American Social Hygiene Association Papers
Association for Voluntary Sterilization (AVS) Papers
National Council on Family Relations Records

PASADENA: CALIFORNIA INSTITUTE OF TECHNOLOGY ARCHIVES

E. S. Gosney Papers and the Records of the Human Betterment Foundation

PHILADELPHIA: AMERICAN PHILOSOPHICAL SOCIETY

American Eugenics Society (AES) Papers
Davenport Papers
Raymond Pearl Papers

SACRAMENTO: CALIFORNIA STATE ARCHIVES

Sonoma State Home Records

STANFORD: STANFORD UNIVERSITY LIBRARY, DEPARTMENT OF SPECIAL COLLECTIONS

David Starr Jordan Papers
Lewis Terman Papers

Annual Reports and Government Documents

California. Department of Institutions. *Biennial Report,* 1922–32.
California. Department of Institutions. *Statistical Report,* 1934–45.
California. Department of Mental Hygiene. *Statistical Report,* 1946–52.
California. State Board of Charities and Corrections. *Biennial Report,*
    1908–22.
California. State Commission in Lunacy. *Biennial Report,* 1904–20.
California Home for the Care and Training of Feeble-Minded Children. *Circular of Information.* Sacramento: California State Printing Office, 1887.
California State Board of Charities and Corrections. *Surveys in Mental Deviation.* Sacramento: State Printing Office, 1918.
Pond, Esther, and Stuart Brody. *Evolution of Treatment Methods at a Hospital for the Mentally Retarded.* California Mental Health Research Monograph 3.
    Sacramento: California Department of Mental Hygiene, 1965.

Books, Articles, and Unpublished Studies

Adams, Mark B., ed. *The Wellborn Science: Eugenics in Germany, France, Brazil, and Russia.* Oxford: Oxford University Press, 1990.
Alexander, Ruth M. *The "Girl Problem": Female Sexual Delinquency in New York, 1900–1930.* Ithaca, N.Y.: Cornell University Press, 1995.
Allen, Frederick Lewis. *Only Yesterday: An Informal History of the 1920s.* New York: Harper & Row, 1964.
Anderson, Karen. *Wartime Women: Sex Roles, Family Relations, and the Status of Women during World War II.* Westport, Conn.: Greenwood Press, 1981.
Armstrong, Elizabeth N. "Hercules and the Muses: Public Art at the Fair." In *The Anthropology of World's Fairs: San Francisco's Panama Pacific International Exposition of 1915,* edited by Burton Benedict. Berkeley, Calif.: Scholar Press, 1983.
Baynton, Douglas. *Forbidden Signs: American Culture and the Campaign against Sign Language.* Chicago: University of Chicago Press, 1996.
Beard, George M. *American Nervousness: Its Causes and Consequences.* New York: Putnam, 1881.
Bederman, Gail. *Manliness and Civilization: A Cultural History of Gender and Race in the United States, 1880–1917.* Chicago: University of Chicago Press, 1995.
Bird, Caroline. *The Invisible Scar.* New York: McKay, 1966.
Brandt, Allan. *No Magic Bullet: A Social History of Venereal Disease in the United States since 1880.* New York: Oxford University Press, 1987.
Braslow, Joel. *Mental Ills and Bodily Cures: Psychiatric Treatment in the First Half of the Twentieth Century.* Berkeley: University of California Press, 1997.
Breines, Wini. *Young, White, and Miserable: Growing Up Female in the Fifties.* Boston: Beacon Press, 1992.
Briggs, Laura. "Discourses of 'Forced Sterilization' in Puerto Rico: The Problem with the Speaking Subaltern." *Differences* 10, no. 2 (1998): 30–66.

Brinkley, Alan. *The Unfinished Nation: A Concise History of the American People.* New York: McGraw-Hill, 1993.

Broberg, Gunnar, and Nils Roll-Hansen, eds. *Eugenics and the Welfare State: Sterilization Policy in Denmark, Sweden, Norway, and Finland.* East Lansing: Michigan State University Press, 1996.

Brown, Calvin. *Official Proceedings of the Second National Conference on Race Betterment.* Battle Creek, Mich.: Race Betterment Foundation, [1916?].

Brown, JoAnne. *The Definition of a Profession: The Authority of Metaphor in the History of Intelligence Testing, 1890–1930.* Princeton, N.J.: Princeton University Press, 1992.

Brumberg, Joan Jacobs. *The Body Project: An Intimate History of American Girls.* New York: Random House, 1997.

Burnham, John C. "The Progressive Era Revolution in American Attitudes toward Sex." *Journal of American History* 59 (1972): 885–908.

Butler, F. O., and Clarence J. Gamble. "Sterilization in a California School for the Mentally Deficient." *American Journal of Mental Deficiency* 6 (April 1947): 745–47.

Butler, Fred. "The Importance of Out-Patient Clinics in State Institutions." *American Journal of Mental Deficiency* 45 (1940): 78–83.

———. "Selective Sterilization: Discussion." *Journal of Psycho-Asthenics* 35 (1930): 51–67.

"California Civic League Meets in Convention." *California Women's Bulletin* 3, no. 3 (January 1915): 27.

Carby, Hazel. *Reconstructing Womanhood: The Emergence of the Afro-American Woman Novelist.* New York: Oxford University Press, 1987.

Cassedy, James H. *Medicine in America.* Baltimore: Johns Hopkins University Press, 1991.

Cawelti, John G. *Apostles of the Self-Made Man.* Chicago: University of Chicago Press, 1965.

Chan, Constance M. *"The Sex Side of Life": Mary Ware Dennett's Pioneering Battle for Birth Control and Sex Education.* New York: New Press, 1996.

Chapman, Paul Davis. *Schools as Sorters: Lewis M. Terman, Applied Psychology, and the Intelligence Testing Movement, 1890–1930.* New York: New York University Press, 1988.

Chauncey, George, Jr. "From Sexual Inversion to Homosexuality: The Changing Medical Conceptualization of Female Deviance." In *Passion and Power: Sexuality in History,* edited by Christina Simmons and Kathy Peiss. Philadelphia: Temple University Press, 1989.

Chesler, Ellen. *Woman of Valor: Margaret Sanger and the Birth Control Movement in America.* New York: Doubleday, 1992.

Cohen, Lizabeth. *Making a New Deal: Industrial Workers in Chicago, 1919–1939.* New York: Cambridge University Press, 1990.

Connelly, Mark. *The Response to Prostitution in the Progressive Era.* Chapel Hill: University of North Carolina Press, 1980.

Coontz, Stephanie. *The Way We Really Are: Coming to Terms with America's Changing Families.* New York: Basic Books, 1997.

Cott, Nancy F. *The Grounding of Modern Feminism.* New Haven, Conn.: Yale University Press, 1987.

——. "Passionlessness: An Interpretation of Victorian Sexual Ideology, 1790–1850." In *A Heritage of Her Own*, edited by Nancy F. Cott and Elizabeth H. Pleck. New York: Simon & Schuster, 1979.

Cravens, Hamilton. *The Triumph of Evolution: American Scientists and the Heredity-Environment Controversy.* Philadelphia: University of Pennsylvania Press, 1978.

Daggett, Mabel Potter. "Women: Building a Better Race." *World's Work* 25 (1912–13): 228–34.

Dally, Ann. *Women under the Knife: A History of Surgery.* New York: Routledge, 1991.

Daniels, Roger. *Coming to America: A History of Immigration and Ethnicity in American Life.* New York: HarperCollins, 1990.

——. *The Politics of Prejudice: The Anti-Japanese Movement in California and the Struggle for Japanese Exclusion.* Berkeley: University of California Press, 1978.

Darwin, Charles. *The Descent of Man.* London: J. Murray, 1871.

Davis, Angela. *Women, Race, and Class.* New York: Vintage Books, 1983.

Davis, Katharine Bement. *Factors in the Sex Life of Twenty-Two Hundred Women.* New York: Harper, 1929.

Dawley, Alan. *Struggles for Justice: Social Responsibility and the Liberal State.* Cambridge, Mass.: Harvard University Press. 1991.

Degler, Carl. *In Search of Human Nature: The Decline and Revival of Darwinism in American Social Thought.* New York: Oxford University Press, 1991.

Dell, Floyd. *Janet March.* New York: Knopf, 1923.

D'Emilio, John, and Estelle Freedman. *Intimate Matters: A History of Sexuality in America.* New York: Harper & Row, 1988.

Deutsch, Albert. "Danger! Venereal Disease." *Nation* 161 (September 22, 1945): 284–85.

Deverell, William, and Tom Sitton, eds. *California Progressivism Revisited.* Berkeley: University of California Press, 1994.

Dickinson, Robert. "Simple Sterilization of Women by Cautery Stricture of Intra-uterine Tubal Openings." *Surgery, Gynecology and Obstetrics* 23 (August 1916): 203–14.

——. "Sterilization without Unsexing: A Surgical Review, with Especial Reference to 5,820 Operations on Insane and Feebleminded in California." In *Collected Papers on Eugenic Sterilization in California: A Critical Study of Results in 6000 Cases.* Pasadena, Calif.: Human Betterment Foundation, 1930.

Dickinson, Robert, and Lura Beam. *The Single Woman: A Medical Study in Sex Education.* New York: Reynal & Hitchcock, 1934.

Dikötter, Frank. "Race Culture: Recent Perspectives on the History of Eugenics." *American Historical Review* 103, no. 2 (April 1998): 467–78.

Doll, Edgar. "Current Thoughts on Mental Deficiency." *Journal of Psycho-Asthenics* 41 (June 1936): 33–49.

Dowbiggin, Ian Robert. *Keeping America Sane: Psychiatry and Eugenics in the United States and Canada, 1880–1940.* Ithaca, N.Y.: Cornell University Press, 1997.

Ellis, Havelock. *Psychology of Sex.* London: Heinemann, 1946.

Emerson, Haven. "From Promoters to Parents: What Can We Expect from the White House Conference?" *Survey* 67 (November 1931): 191–92.

"Eugenical Sterilization at the Meeting of the AMA." *Eugenical News* 13 (August 1928): 115.

Faludi, Susan. *Backlash: The Undeclared War against American Women.* New York: Crown, 1991.

Fass, Paula. *The Damned and the Beautiful: American Youth in the 1920s.* New York: Oxford University Press, 1977.

Fernald, Walter E. *The Burden of Feeble-mindedness.* Boston: Massachusetts Society for Mental Hygiene, 1918.

———. "Care of the Feeble-minded." In *Proceedings, National Conference of Charities and Corrections.* Indianapolis: Press of Frederick J. Heer, 1904.

Filene, Peter. *Him/Her/Self: Sex Roles in Modern America.* 2d ed. Baltimore: Johns Hopkins University Press, 1986.

Fishbein, Morris, ed. *Successful Marriage: A Modern Guide to Love, Sex, and Family Life.* Garden City, N.Y.: Doubleday, 1963.

Fox, Richard. *So Far Disordered in Mind: Insanity in California, 1870–1930.* Berkeley: University of California Press, 1978.

Gilman, Charlotte Perkins. "Progress through Birth Control." *North American Review* 224 (December 1927): 622–29.

———. *Women and Economics.* Boston: Small, Maynard, 1899.

Gilman, Sander. *Difference and Pathology: Stereotypes of Sexuality, Race, and Madness.* Ithaca, N.Y.: Cornell University Press, 1985.

Goddard, Henry H. "Four Hundred Feeble-minded Children Classified by the Binet Method." *Journal of Genetic Psychology* 17, no. 3 (1910): 387–97.

———. *The Menace of Mental Deficiency from the Standpoint of Heredity.* Vineland: New Jersey Training School, 1915.

Gordon, Linda. *Heroes of Their Own Lives: The Politics and History of Family Violence.* New York: Penguin Books, 1988.

———. *Woman's Body, Woman's Right: Birth Control in America.* New York: Penguin Books, 1990.

Gorn, Elliott J. *The Manly Art: Bare-Knuckle Prize Fighting in America.* Ithaca, N.Y.: Cornell University Press, 1986.

Gosney, E. S., and Paul Popenoe. *Sterilization for Human Betterment: A Summary of Results of 6,000 Operations in California, 1909–1929.* New York: Macmillan, 1929.

———. *Twenty-Eight Years of Sterilization in California.* Pasadena, Calif.: Human Betterment Foundation, 1938.

Gould, Stephen Jay. *The Mismeasure of Man.* New York: Norton, 1981.

Graham, Sylvester. *Chastity, in a Course of Lectures to Young Men: Intended Also for the Serious Consideration of Parents and Guardians.* New York: Fowler and Wells, n.d.

Grether, Judith K. "Sterilization and Eugenics: An Examination of Early Twentieth Century Population Control in the United States." Ph.D. diss., University of Oregon, 1980.

Griswold, Robert. *Fatherhood in America: A History.* New York: Basic Books, 1993.

Groneman, Carol. "Nymphomania: The Historical Construction of Female Sexuality." *Signs,* Winter 1994, 337–67.

Grossman, Atina. *Reforming Sex: The German Movement for Birth Control and Abortion Reform, 1920–1950.* New York: Oxford University Press, 1995.

Hall, Jacquelyn Dowd. *Revolt against Chivalry: Jessie Daniel Ames and the Women's Campaign against Lynching.* New York: Columbia University Press, 1974.

Haller, Mark. *Eugenics: Hereditarian Attitudes in American Thought.* New Brunswick, N.J.: Rutgers University Press, 1963.

Hay, James, Jr. "'A Wise Son Maketh a Glad Father': The President's November Conference on Child Health Will Work to Develop Our Greatest Asset," *World's Work* 59 (October 1930): 70–73.

Higham, John. *Strangers in the Land: Patterns of American Nativism 1860–1925.* New York: Atheneum, 1971.

Hinkle, Beatrice. "Women and the New Morality." In *Our Changing Morality: A Symposium,* edited by Freda Kirchway. New York: Boni, 1924.

Hirsch, James S. "Happy Endings: This Marriage, Too, Is Saved by Therapy." *Wall Street Journal,* 14 January 1992, p. A1.

Hoffman, Betty Hannah. "The Man Who Saves Marriages." *Ladies' Home Journal,* September 1960, 71, 120, 123–25.

Holt, William L. *The Venereal Peril.* New York: Eugenics Publishing, 1939.

Horn, David G. "This Norm Which Is Not One: Reading the Female Body in Lombroso's Anthropology." In *Deviant Bodies: Critical Perspectives on Difference in Science and Popular Culture,* edited by Jennifer Terry and Jacqueline Urla. Bloomington: Indiana University Press, 1995.

Huntington, Ellsworth. *Tomorrow's Children: The Goal of Eugenics.* New York: Wiley, 1935.

Institute for American Values. "Introduction and Overview," http://www.americanvalues.org.

Jacoby, Russell, and Naomi Glauberman. *The Bell Curve Debate: History, Documents, Opinions.* New York: Random House, 1995.

Johnson, Bascom. "Next Steps." *Journal of Social Hygiene* 4 (1918): 9–23.

Kessler-Harris, Alice. *Out to Work: A History of Wage-Earning Women in the United States.* New York: Oxford University Press, 1982.

Kevles, Daniel J. *In the Name of Eugenics: Genetics and the Uses of Human Heredity.* Berkeley: University of California Press, 1985.

Kimmel, Michael. *Manhood in America: A Cultural History.* New York: Free Press, 1996.

Kline, Wendy. "'Building a Better Race': Eugenics and the Making of Modern Morality in America, 1900–1960." Ph.D. diss., University of California at Davis, 1998.

Kühl, Stefan. *The Nazi Connection: Eugenics, American Racism, and German National Socialism.* New York: Oxford University Press, 1994.

Kunzel, Regina. *Fallen Women, Problem Girls: Unmarried Mothers and the Professionalization of Social Work, 1890–1945.* New Haven, Conn.: Yale University Press, 1993.

Ladd-Taylor, Molly. *Mother Work: Women, Child Welfare, and the State.* Urbana: University of Illinois Press, 1994.

Landman, J. H. *Human Sterilization: The History of the Sexual Sterilization Movement.* New York: Macmillan, 1932.

Lane, Ann J. Introduction to *Herland,* by Charlotte Perkins Gilman. New York: Pantheon Books, 1979; *Herland* originally published 1915.

Larson, Edward J. *Sex, Race, and Science: Eugenics in the Deep South.* Baltimore: Johns Hopkins University Press, 1995.

Lasch, Christopher. *Haven in a Heartless World: The Family Besieged.* New York: Norton, 1977.

Lears, T. J. Jackson. *No Place of Grace: Antimodernism and the Tranformation of American Culture, 1880–1920.* New York: Pantheon Books, 1981.

Leavenworth, Isabel. "Virtue for Women." In *Our Changing Morality: A Symposium,* edited by Freda Kirchway. New York: Boni, 1924.

Leavitt, Judith Walzer. *Typhoid Mary: Captive to the Public's Health.* Boston: Beacon Press, 1996.

Lederer, Susan. *Subjected to Science: Human Experimentation in America before the Second World War.* Baltimore: Johns Hopkins University Press, 1995.

Lunbeck, Elizabeth. "'A New Generation of Women': Progressive Psychiatrists and the Hypersexual Female." *Feminist Studies* 13, no. 3 (Fall 1987): 513–43.

——— . *The Psychiatric Persuasion: Knowledge, Gender, and Power in Modern America.* Princeton, N.J.: Princeton University Press, 1994.

Maines, Rachel. *The Technology of Orgasm: "Hysteria," the Vibrator, and Women's Sexual Satisfaction.* Baltimore: Johns Hopkins University Press, 1999.

Marchand, Roland. *Advertising the American Dream: Making Way for Modernity, 1920–1940.* Berkeley: University of California Press, 1985.

Marsh, Margaret, and Wanda Rommer. *The Empty Cradle: Infertility in America from Colonial Times to the Present.* Baltimore: Johns Hopkins University Press, 1996.

Massachusetts League for Preventive Work. *Feeble-minded Adrift: Reasons Why Massachusetts Needs a Third School for the Feeble-minded IMMEDIATELY.* Boston: n.p., 1916.

May, Elaine. *Barren in the Promised Land: Childless Americans and the Pursuit of Happiness.* New York: Basic Books, 1995.

——— . *Great Expectations: Marriage and Divorce in Post-Victorian America.* Chicago: University of Chicago Press, 1980.

——— . *Homeward Bound: American Families in the Cold War Era.* New York: Basic Books, 1988.

McCann, Carole R. *Birth Control Politics in the United States, 1916–1945.* Ithaca, N.Y.: Cornell University Press, 1994.

McGovern, James. "The American Woman's Pre–World War I Freedom in Manners and Morals." *Journal of American History* 55 (1968): 315–33.

McLaughlin, A. J. "Pioneering in Venereal Disease Control." *American Journal of Obstetrics* 80 (December 1919).

Mehler, Barry Alan. "A History of the American Eugenics Society, 1921–1940." Ph.D. diss., University of Illinois at Urbana-Champaign, 1988.

Meyerowitz, Joanne J. *Women Adrift: Independent Wage Earners in Chicago, 1880–1930.* Chicago: University of Chicago Press, 1988.

Minton, Henry L. *Lewis M. Terman: Pioneer in Psychological Testing.* New York: New York University Press, 1988.

Mintz, Steven, and Susan Kellogg. *Domestic Revolutions: A Social History of American Family Life.* New York: Free Press, 1988.

Modell, John. *Into One's Own: From Youth to Adulthood in the United States, 1920 to 1975*. Berkeley: University of California Press, 1989.

Morantz-Sanchez, Regina. *Conduct Unbecoming a Woman: Medicine on Trial in Turn-of-the-Century Brooklyn*. New York: Oxford University Press, 1999.

Noll, Steven. *Feeble-minded in Our Midst: Institutions for the Mentally Retarded in the South, 1900–1940*. Chapel Hill: University of North Carolina Press, 1995.

Odem, Mary E. *Delinquent Daughters: Protecting and Policing Adolescent Female Sexuality in the United States, 1885–1920*. Chapel Hill: University of North Carolina Press, 1995.

Parenthood. 124 min. Universal Pictures, Universal City, Calif., 1989.

Patterson, James T. *Grand Expectations: The United States, 1945–1974*. New York: Oxford University Press, 1996.

Paul, Diane. *Controlling Human Heredity, 1865 to the Present*. Atlantic Highlands, N.J.: Humanities Press, 1995.

Peiss, Kathy. "'Charity Girls' and City Pleasures: Historical Notes on Working-Class Sexuality, 1880–1920." In *Passion and Power: Sexuality in History*, edited by Christina Simmons and Kathy Peiss. Philadelphia: Temple University Press, 1989.

——. *Cheap Amusements: Working Women and Leisure in Turn-of-the-Century New York*. Philadelphia: Temple University Press, 1986.

Pernick, Martin. *The Black Stork: Eugenics and the Death of "Defective" Babies in American Medicine and Motion Pictures since 1915*. New York: Oxford University Press, 1996.

Pickens, Donald K. *Eugenics and the Progressives*. Nashville, Tenn.: Vanderbilt University Press, 1968.

Popenoe, David. *Life without Father: Compelling New Evidence That Fatherhood and Marriage Are Indispensable for the Good of Children and Society*. Cambridge, Mass.: Harvard University Press, 1996.

Popenoe, Paul. "Eugenic Sterilization in California: Effect of Salpingectomy on the Sexual Life." In *Collected Papers on Eugenic Sterilization in California: A Critical Study of Results in 6,000 Cases*. Pasadena, Calif.: Human Betterment Foundation, 1930.

——. "Eugenic Sterilization in California: The Feebleminded." In *Collected Papers on Eugenic Sterilization in California: A Critical Study of Results in 6,000 Cases*. Pasadena, Calif.: Human Betterment Foundation, 1930.

——. "The German Sterilization Law." *Journal of Heredity* 25, no. 7 (July 1934): 257–60.

——. "How Can Colleges Prepare Their Students for Marriage and Parenthood?" *Journal of Home Economics* 22, no. 3 (March 1930): 169–78.

——. "Success on Parole after Sterilization." In *Collected Papers on Eugenic Sterilization in California: A Critical Study of Results in 6000 Cases*. Pasadena, Calif: Human Betterment Foundation, 1930.

——. "The Writings of Dr. Popenoe." *Family Life: Special Memorial Issue* 39, no. 5 (September and October 1979): 9–16.

Popenoe, Paul, and Norman Fenton. "Sterilization as a Social Measure." *Journal of Psycho-Asthenics* 41 (1936): 60–65.

Popenoe, Paul, and Roswell Hill Johnson. *Applied Eugenics*. Rev. ed. New York: Macmillan, 1933.

Prouty, Olive Higgins. *Stella Dallas*. New York: Grosset & Dunlap, 1923.

Reed, James. *From Private Vice to Public Virtue: The Birth Control Movement and American Society since 1830*. Princeton, N.J.: Princeton University Press, 1978.

Reilly, Philip. *The Surgical Solution: A History of Involuntary Sterilization in the United States*. Baltimore: Johns Hopkins University Press, 1991.

Review of *Janet March*, by Floyd Dell. *Bookman* 58 (December 1923): 453.

Robinson, Paul. *The Modernization of Sex: Havelock Ellis, Alfred Kinsey, William Masters and Virginia Johnson*. New York: Harper & Row, 1976.

Rockafellar, Nancy. "Making the World Safe for the Soldiers of Democracy: Patriotism, Public Health and Venereal Disesase Control on the West Coast, 1910–1919." PhD. diss., University of Washington, 1990.

Rodriguez-Trìas, Helen. "Sterilization Abuse." In *Biological Woman—the Convenient Myth: A Collection of Feminist Essays and a Comprehensive Bibliography*. Cambridge, Mass.: Schenkman Books, 1982.

Rogers, Anna A. "Why American Marriages Fail." *Atlantic Monthly* 100 (September 1907): 289–98.

Rogers, Daniel. "In Search of Progressivism." *Reviews in American History* 10 (December 1982): 113–32.

Rosen, Margery, and the Editors of the *Ladies' Home Journal. Can This Marriage Be Saved? Real-Life Cases from the Most Popular, Most Enduring Women's Magazine Feature in the World*. New York: Workman, 1994.

Rosen, Ruth. *The Lost Sisterhood: Prostitution in America, 1900–1918*. Baltimore: Johns Hopkins University Press, 1982.

Ross, Dorothy. *G. Stanley Hall: The Psychologist as Prophet*. Chicago: University of Chicago Press, 1972.

Rothman, David J. *The Discovery of the Asylum: Social Order and Disorder in the New Republic*. Boston: Little, Brown, 1971.

Russett, Cynthia Eagle. *Sexual Science: The Victorian Construction of Womanhood*. Cambridge, Mass.: Harvard University Press, 1989.

Rydell, Robert. *World of Fairs: The Century-of-Progress Expositions*. Chicago: University of Chicago Press, 1993.

Scanlon, Jennifer. *Inarticulate Longings: The Ladies' Home Journal, Gender, and the Promises of Consumer Culture*. New York: Routledge, 1995.

Schoen, Johanna. "'A Great Thing for Poor Folks': Birth Control, Sterilization, and Abortion in Public Health and Welfare in the Twentieth Century." Ph.D. diss., University of North Carolina, Chapel Hill, 1995.

Scull, Andrew, and Diane Favreau. "'A Chance to Cut Is a Chance to Cure': Sexual Surgery for Psychosis in Three Nineteenth Century Cities." *Research in Law, Deviance and Social Control* 8 (1986): 3–39.

Seabury, Florence Guy. "Stereotypes." In *Our Changing Morality: A Symposium*, edited by Freda Kirchway. New York: Boni, 1924.

Simmons, Christina. "Companionate Marriage and the Lesbian Threat." In *Women and Power in American History: A Reader*, vol. 2, edited by Kathryn Kish Sklar and Thomas Dublin. Englewood Cliffs, N.J.: Prentice-Hall, 1991.

————. "Modern Sexuality and the Myth of Victorian Repression." In *Passion and Power: Sexuality in History,* edited by Christina Simmons and Kathy Peiss. Philadelphia: Temple University Press, 1989.

Smith-Rosenberg, Carroll. "Discourses of Sexuality and Subjectivity: The New Woman, 1870–1936." In *Hidden from History: Reclaiming the Gay and Lesbian Past,* edited by Martin Duberman, Martha Vicinus, and George Chauncey Jr. New York: New American Library, 1989.

Solinger, Rickie. *Wake Up Little Susie: Single Pregnancy and Race before Roe v. Wade.* New York: Routledge, 1992.

Soloway, Richard A. "The 'Perfect Contraceptive': Eugenics and Birth Control Research in Britain and America in the Interwar Years." *Journal of Contemporary History* 30 (1995): 636–64.

Stacey, Judith. *In the Name of the Family: Rethinking Family Values in the Postmodern Age.* Boston: Beacon Press, 1996.

Starr, George. "Truth Unveiled: The Panama Pacific International Exposition and Its Interpreters." In *The Anthropology of World's Fairs: San Francisco's Panama Pacific International Exposition of 1915,* edited by Burton Benedict. Berkeley, Calif.: Scholar Press, 1983.

Stepan, Nancy Leys. *"The Hour of Eugenics": Race, Gender, and Nation in Latin America.* Ithaca, N.Y.: Cornell University Press, 1996.

Stoddard, Lothrop. *The Revolt against Civilization.* New York: Scribner, 1923.

Terman, Lewis. *Psychological Factors in Marital Happiness.* New York: McGraw-Hill, 1938.

Terman, Lewis, and Catherine Cox Miles. *Sex and Personality: Studies in Masculinity and Femininity.* New York: Russell & Russell, 1936.

Terry, Jennifer. "Anxious Slippages between 'Us' and 'Them': A Brief History of the Scientific Search for Homosexual Bodies." In *Deviant Bodies: Critical Perspectives on Difference in Science and Popular Culture,* edited by Jennifer Terry and Jacqueline Urla. Bloomington: Indiana University Press, 1995.

Theriot, Nancy. *Mothers and Daughters in Nineteenth-Century America: The Biosocial Construction of Femininity.* Lexington: University Press of Kentucky, 1996.

Thomson, Rosemarie. *Extraordinary Bodies: Figuring Physical Disability in American Culture and Literature.* New York: Columbia University Press, 1997.

Trachtenberg, Alan. *The Incorporation of America: Culture and Society in the Gilded Age.* New York: Hill & Wang, 1982.

Trent, James W. *Inventing the Feeble Mind: A History of Mental Retardation in the United States.* Berkeley: University of California Press, 1994.

Trimberger, Ellen Kay. "Feminism, Men, and Modern Love: Greenwich Village, 1900–1925." In *Powers of Desire: The Politics of Sexuality,* edited by Ann Snitow, Christine Stansell, and Sharon Thompson. New York: Monthly Review Press, 1983.

Trombley, Stephen. *The Right to Reproduce: A History of Coercive Sterilization.* London: Weidenfeld and Nicholson, 1988.

Urla, Jacqueline, and Jennifer Terry. "Introduction: Mapping Embodied Deviance." In *Deviant Bodies: Critical Perspectives on Difference in Science and*

Popular Culture, edited by Jennifer Terry and Jacqueline Urla. Bloomington: Indiana University Press, 1995.

VanEssendelft, William. "A History of the Association for Voluntary Sterilization: 1935–1964." Ph.D. diss., University of Minnesota, 1978.

Van Horn, Susan Householder. *Women, Work, and Fertility.* New York: New York University Press, 1988.

Walkowitz, Judith. *City of Dreadful Delight: Narratives of Sexual Danger in Late-Victorian London.* Chicago: University of Chicago Press, 1992.

Walters, Ronald G. *Primers for Prudery: Sexual Advice to Victorian America.* Englewood Cliffs, N.J.: Prentice-Hall, 1974.

Weeks, Jeffrey. *Sexuality and Its Discontents: Meanings, Myths & Modern Sexualities.* London: Routledge, 1985.

Welter, Barbara. "The Cult of True Womanhood, 1820–1860." *American Quarterly* 18 (1966): 131–75.

Westbrook, Robert B. "Fighting for the American Family: Private Interests and Political Obligation in World War II." In *The Power of Culture: Critical Essays in American History,* edited by Richard Fox and T. J. Jackson Lears. Chicago: University of Chicago Press, 1993.

Wiebe, Robert. *The Search for Order, 1877–1920.* New York: Hill & Wang, 1967.

Wilbur, Ray Lyman. "Toward a Better Child Life." *Review of Reviews* 83 (January 1931): 55.

Woloch, Nancy. *Women and the American Experience.* New York: McGraw-Hill, 1992.

Zenderland, Leila. "The Debate over Diagnosis: Henry Herbert Goddard and the Medical Acceptance of Intelligence Testing." In *Psychological Testing and American Society, 1880–1930,* edited by Michael M. Sokal. New Brunswick, N.J.: Rutgers University Press, 1987.

Zunz, Olivier. *Making America Corporate 1870–1920.* Chicago: University of Chicago Press, 1990.

Index

Numbers in italics indicate illustrations.

Text: 10/13 ITC Galliard
Display: Galliard
Compositor: Impressions Book & Journal Services, Inc.
Printer & Binder: Sheridan Books, Inc.